LORD CURZON
The Last of The British Moghuls

LORD CURZON

The Last of The British Moghuls

Nayana Goradia

DELHI
OXFORD UNIVERSITY PRESS
BOMBAY CALCUTTA MADRAS

Oxford University Press, Walton Street, Oxford OX2 6DP

Oxford New York Toronto
Delhi Bombay Calcutta Madras Karachi
Kuala Lumpur Singapore Hong Kong Tokyo
Nairobi Dar es Salaam Cape Town
Melbourne Auckland Madrid

and associates in
Berlin Ibadan

First published 1993
Third impression 1994

ISBN 0 19 562824 1

Endpapers: Government House, Calcutta, 1876.
Courtesy Illustrated London News Picture Library.
Title page: Victoria Memorial, Calcutta
Drawing by Vishwajyoti Ghosh.

Typeset by Imprinter, C-79 Okhla Phase-I, New Delhi 110020
Printed by Rekha Printers Pvt. Ltd., New Delhi 110020
and published by Neil O'Brien, Oxford University Press
YMCA Library Building, Jai Singh Road, New Delhi 110001

To
Prafull
without whom this
would not have been written, and
Madhavi, Brinda and Gautam

Contents

Contents

List of Illustrations

Prologue

Late in the evening of Monday 21 May 1923, a loud knock disturbed the serene tranquility of Montacute, the stately country retreat in Somerset of Britain's Foreign Secretary Marquess Curzon of Kedleston. The caller was the village policeman who had arrived on his bicycle with an urgent telegram. Being the Whitsuntide recess, the local post-office boy had been on holiday but the importance of the telegram was enough to galvanize the policeman into delivering the message himself. It was a royal summons asking Curzon to return to London at once; the sender, Lord Stamfordham, was Private Secretary to the King of England.

To Lord Curzon this could have only one meaning. At last the years of waiting seemed to be over and he was being summoned as the new Prime Minister of Great Britain. Earlier in the day, Curzon had received a letter from his ailing chief, Prime Minister Bonar Law, saying that illness prevented him from continuing in office. Bonar Law had written: 'I understand that it is not customary for the King to ask the P M to recommend a successor in circumstances like the present and I presume that he will not do so; but if, as I hope, he accepts my resignation at once, he would have to take immediate steps about my successor.'

Curzon had few doubts about his elevation. He was Foreign

Secretary and the leader of the House of Lords. Besides, during Law's increasing absence from the Cabinet due to illness, it was Curzon who had acted as Deputy Prime Minister and presided over its meetings. Few could challenge his seniority, his brilliance, his knowledge and his claim to the highest office in the land.

In his early years destiny seemed to have smiled most generously on Curzon. His family, though not wealthy, traced its descent in an unbroken line to the Norman Conquest. At Eton he had won an extraordinary number of honours and while he was still at Oxford his speeches were being quoted in Parliament. As a young MP he had undertaken journeys to far-flung outposts of the Empire. The books he wrote on his travels earned him the reputation of one of the best informed men on the Empire. At 39 he was Viceroy of India, the youngest man in history to occupy that office.

In 1916 Prime Minister Lloyd George had brought him into the centre of Britain's political arena by appointing him to the crucial four-member War Cabinet. When in 1919 he had been elevated to the Foreign Office, the Near East was in ferment. Peace talks had collapsed and in the crisis of September 1922, Curzon maintained his attempts had averted the disaster at Chanak. Even hostile newspapers had hailed Curzon's role, calling him the 'Peace-Maker' in a 'Cabinet of warmongers'.

When Lloyd George's coalition government collapsed in October the premiership seemed to be within Curzon's grasp. But it was at this precise moment, that elder Tory statesman Bonar Law decided to return to the political field. Curzon had little option but to bide his time. Now, less than seven months later, news came of Law's retirement. It seemed as though destiny was once again on his side.

At 64 Curzon was still a handsome man. The lofty, somewhat disdainful air of the patrician came naturally to him. He held himself erect with a ramrod stiffness. The passing years had done little to breach that supreme self-assuredness which had prompted some mischief makers, while he was still at Oxford, to put into verse a doggerel about him, that went:

> My name is George Nathaniel Curzon,
> I am a most superior person,
> My cheek is pink, my hair is sleek,
> I dine at Blenheim once a week.

Busy, overworked and almost always in constant pain from a childhood injury due to a fall from horseback, Curzon found time to play host to a motley crew of rather distinguished people at his various great houses in London and the country. In congenial company, the arrogant mask fell away; he could be a high-spirited companion.

Women found him irresistible. Grace, the second Lady Curzon, was an exception. She had the good looks of a Grecian goddess but could be selfish and cruel. Eighteen years younger than Curzon, Grace had been the widow of an Argentine millionaire whose estates were so large that he owned his own railroad. Marriage to Curzon gave her the social status which she craved but had hitherto been denied. As wife of a senior British cabinet minister, she was automatically launched as one of the foremost hostesses in the land. This was all she seemed to care for. When he was ill and in pain, she was rarely by his side. Curzon's Papers are full of sad, wistful notes to his wife; 'No promised letter from Grace today and I wonder if she will really come down tomorrow.' She rarely did.

A week before the Whitsuntide recess, Grace had been holidaying by herself in Paris, staying at the glittering Ritz hotel. As the Marchioness Curzon of Kedleston and wife of Britain's Foreign Secretary, she was fêted at the embassies. She dined with the King of Portugal and the Agha Khan sent her flowers. 'If it were not for you', she told her husband as though registering a sacrifice, 'I would much rather spend the season here, as London has grown stale for me.' This was a few days before she received firsthand news of Bonar Law's illness from the British Embassy in Paris, giving rise to talk of his immediate retirement and rumours of Curzon's elevation to premiership.

Bonar Law had been ailing for some time. In early May, on doctor's advice, Law had embarked on a Mediterranean cruise.

But by the time the ship reached Genoa he was in such agony as to be taking several aspirins during the day. He was forced to disembark from the ship and was taken to Paris. There Thomas Horder, a London physician brought over to examine him, pronounced the Prime Minister had cancer of the throat of such malignancy that he could not hope to survive till the end of the year.

Lady Crewe, wife of the British Ambassador in Paris, rushed across to tell Grace the news. Her advice had been that Grace should promptly return to England and be with Curzon, for in the circumstances, as Grace has recorded, both '... she and Lord Crewe felt certain that he [Curzon] would be sent for by the King and was bound to be the next Prime Minister.' While Lady Crewe was with her, Grace put a call to Curzon in London to pass on the news. The two had been scheduled to go to Montacute for the weekend but in view of the new developments Grace thought Curzon might feel it prudent to stay where he was. But Curzon stubbornly stuck to his plan. With his old-world views, he felt it more behoving to go out to his country seat and await there the prize he believed was coming his way.

Grace willingly fell in with his wishes, no longer averse to abandon the charms of Paris for the solitude of the English countryside. This time she was not going to back out at the last minute, as was her habit. Grace sent her husband a loving little note from Paris assuring him, 'I return to my Boy on Thursday and look forward so much to going to Montacute with him on Friday.' She was unable to resist adding, 'If my Boy becomes P M, I will not go to B. A. [Buenos Aires] as he will need me at home.'

Grace dutifully returned from Paris and on Friday morning they left for Montacute. Set in the heart of Somerset, Montacute is one of the few surviving classically-Tudor villas in Britain. Built in the sixteenth century by one of Queen Elizabeth's courtiers, it was designed for royal visits and the Queen is said to have spent a night there. Its great columned halls, filled with priceless furniture and paintings, open to vast courtyards.

Upstairs a long gallery branches off into the bed-chambers. As with all the Curzon country homes, Montacute was fully equipped for great house parties. Each establishment boasted a permanent house-keeper with a retinue of maids. The men servants travelled with the Curzons in order to keep them in princely comfort.

The only comfort Montacute seemed to be lacking, however, was a telephone. Curzon later said, 'Not being on the telephone, I could not communicate with anyone in London, and I naturally refrained from hurrying back, lest any action should be misrepresented.' So he sat and waited in remote Montacute while his detractors had a field day in London.

On Monday morning Curzon received Bonar Law's letter about his resignation and his warning that he did not intend to nominate a successor. That Whit Monday was a day of 'intense anxiety, speculation and hope for George and me', Grace was to recall. 'George feeling restless and excited kept on coming in from the garden and talking to me about the political future.' Still he made no plans to return to London. The moment of supreme triumph—the moment for which he had geared his life—was coming his way. Yet he kept away; desperately wanting to go, but determined not to until he was called. He was throwing a gauntlet to destiny but such was his narcissism that he could not act otherwise. The long agonizing wait ended in the evening when the royal summons came.

By next morning the news had spread over Yeovil, the sleepy little town near Montacute. From the early hours, a steady trickle of people started collecting at the railway station with flags and bunting. When at last the Curzons drove onto the platform, they surged forward, cheering and clapping. Grace said:

> My husband was full of hope and the most buoyant optimism as we travelled from Yeovil to Paddington. Our journey from Montacute to 1 Carlton House Terrace must have been the most photographed one of our lives together. All newspapers seemed to be sure that George was about to become Prime Minister.

Grace had a severe toothache which had given her a swollen cheek, which she had camouflaged with a veil. But flushed with elation she did not mind the reporters and the photographers. The pain did not prevent her from making plans about the rosy future opening up before them. She assured 'her Boy' that she would be by his side at the countless receptions and balls they would be obliged to give. Delighted to find her responsive, Curzon fell in with her plans. They decided that they would use the Prime Minister's residence, No. 10 Downing Street, only for official functions. All their private entertaining would be at their own home, 1 Carlton House Terrace.

Designed in the Regency style and only a short distance from Buckingham Palace, 1 Carlton House Terrace is situated in London's most expensive and fashionable district. Its tall windows open on The Mall where the great London processions pass. The fine halls, the upstairs ballroom and grand staircase made it ideal for receptions.

In the train Curzon had glanced at the newspapers. They too seemed to be overwhelmingly on his side.

Curzon later said:

> On my journey to London, I had found in the morning Press an almost unanimous opinion that, the choice lying between Baldwin and myself, there was no question as to the immense superiority of my claims and little doubts as to the intentions of the King. The crowd of Press photographers at Paddington and my house—deceptive and worthless as these phenomena are—at least indicated that popular belief.

The Curzons returned to London reaching Carlton House Terrace at 1.20 p.m. in time for lunch. A message from Lord Stamfordham saying that he would call at 2.30 p.m. the same afternoon was waiting for Curzon. Lord Stamfordham arrived on the dot of the hour and was ushered into a large room overlooking the trees of St. James Park. Curzon insisted that Grace be present at the meeting. Curzon recorded the interview:

> Lord Stamfordham began by explaining...the great difficulty which the King had been placed in. H M had been much offended at not being apprised of Bonar Law's hasty return to England, nor of his intention to resign, nor of his resignation itself, which he had first seen in the papers at Aldershot before the P M's letter had been

placed in his hands. He [the King] had further failed to receive, though he had sought it, any advice from Bonar Law as to his successor. In these circumstances, the King had to consult such opinion as he could, and Lord Stamfordham had been sent to London on the previous day for the purpose, and had seen Lord Salisbury and some other unnamed politicians.

Curzon understood that these were formalities that any monarch in the circumstances would be obliged to complete. But he felt a wave of irritation at Bonar Law for putting the King into such a predicament.

Next Lord Stamfordham went on to explain, the King fully recognized my pre-eminent claims—immeasurably superior, as he expressed it, to those of any other candidate. These claims were, he said, such as to entitle me to expect the succession for which the King regarded me as in every way qualified.

Though the War had ended, crucial negotiations were pending before actual peace could return to Europe. At the peace talks at Lausanne, Curzon had earned the reputation of a successful negotiator. The previous September, when Turkey had risen up in arms, his skills had helped to bring the warring sides to the negotiating table. The Greeks, having gone on the offensive in Asia Minor, with Lloyd George's encouragement, had been confronted with the wrath of the Turks. Uniting under Kemal Pasha, the Turkish army burnt down Smyrna, drove the Greeks into the sea and began casting eyes on the Allied Neutral Zone at Chanak. Curzon had then happily said to his wife, 'I wonder if you read the wonderful tributes about me in the English newspapers...I have suddenly been discovered at the age of 63. I was discovered when I was Viceroy of India from 39–46, then I was forgotten, traduced, buried, ignored. Now I have been dug up again...' But the resurrection was shortlived.

...now the blow fell—the King had convinced himself that, inasmuch as the largest section of the Opposition to the present Government was a Labour Opposition and as that party was not represented in the House of Lords, the future Leader of the

Government must be in the House of Commons; where he could answer the Labour Leader with full authority. Accordingly the King had decided to pass over and appoint Stanley Baldwin instead.

Curzon was stunned. He could scarcely believe his ears. Tears of bewilderment and shock rushed to his eyes. Holding himself together, Curzon thanked the King for his compliments and for the honour of sending his personal emissary. He also pointed out that the King's decision was tantamount to laying down a doctrine that a peer could never again be PM. 'I earnestly protest against the new doctrine,' he said. 'It involves an additional and perpetual and cruel disability upon the order of the House to which I belong.'

But protest was futile. The tears could no longer be held back. He had lost the race. What greater humiliation was there for him than to be passed over by Stanley Baldwin whom he had looked upon as 'a man of utmost insignificance'? The proud Lord Curzon is said to have broken down and wept like a child.

Son of a Worcestershire iron magnate, Stanley Baldwin had secured a poor second at Cambridge and was set for the humdrum life of the director of the family iron works had it not been for his father's premature death in 1908. Baldwin was chosen to take his place as Conservative MP for Bewdley. From then onward, he began cultivating the role of a dull but thoroughly trustworthy English country gentleman. Sensing that this was what the people of post-war Britain wanted, Baldwin put across the message that the dangerous days of the brilliant, dynamic wartime leaders like David Lloyd George and Churchill were over. In the Cabinet crisis of October 1922, Stanley Baldwin had provoked the revolt of junior Tory ministers which culminated in Lloyd George's overthrow.

In the race for political promotion, Curzon had been miles ahead. He was Viceroy in 1899 whereas Baldwin only entered Parliament at the age of 43 in 1908. When Curzon was a part of the War Cabinet, Baldwin made his début as junior minister. With Bonar Law's encouragement, however, Baldwin had steadily risen to become, within six years, the Leader of the House of Commons and Chancellor of the Exchequer. But rooted as he was in a bygone era of protocol and privilege, Curzon could never quite overcome the assumption that the art of governing was the prerogative of men born to do so. That is why, perhaps, he had not felt unduly threatened by the 58 year-old Stanley Baldwin. Ironically, Curzon was not unaware of Bonar Law's partiality for Baldwin. Only a month before the resignation, he had written to Law categorically stating, 'I have been asked three times within the last three weeks whether it is true that you are about to resign and recommend that Baldwin be appointed in your place?' It was not a letter calculated to endear him to Law.

A pall of gloom hung over the air that weekend at 4 Onslow Gardens, Bonar Law's London residence. Law had returned to London on Saturday. Wise diplomacy, if not human consideration, required that Curzon stay back in London to receive his sick chief. Neither factor had weighed heavily with him when he left for Montacute on his usual holiday. Stanley Baldwin, on the other hand, had remained present over that entire weekend to receive Law and had never moved far from Onslow Gardens.

Bonar Law's Parliamentary Private Secretary, Lord Davidson, who was with him in Paris, had returned earlier to London to inform the Cabinet and the King of his illness. As a special friend of Baldwin, Davidson had taken pains to keep him in the picture. Baldwin had promptly driven down to London and the two men dined together at the Argentine Club. Davidson apparently told Baldwin of Law's decision of not naming a successor and also made it clear that he would prefer Baldwin to be the next Prime Minister.

At 10.30 on Sunday morning Baldwin called on Bonar Law. The Prime Minister led him to understand that though he did not intend to recommend a successor, he had no doubt that Curzon would be the automatic choice. Apparently Law went on to assure Baldwin that his chance was bound to come in due course, and the delay might even be to his advantage since it would give him the opportunity of gaining experience in the management of the House of Commons. Meanwhile, he must try to serve Curzon loyally. Baldwin is said to have replied that he would gladly serve under anyone who could hold the party together.

Bonar Law next made preparations to reach King George V who was thirty-seven miles out of London, relaxing in the Royal Pavilion at Aldershot. Too ill to go himself, he assigned his Principal Private Secretary, Colonel Ronald Waterhouse, with the task of carrying his resignation letter. Waterhouse was accompanied by Bonar Law's son-in-law, Sir Frederick Sykes. Accordingly, on Sunday afternoon Waterhouse and Sykes drove by motor car to Aldershot with Law's resignation letter. They were received at the Royal Pavilion where the letter was delivered to the King.

As Bonar Law had not mentioned his successor, the King sent his Secretary, Lord Stamfordham, to London to consult senior Tory members. On Monday morning Lord Stamfordham drove to London. He met several elder Tory statesmen, including Arthur Balfour and Lord Salisbury. After consultations, a telegram was despatched to Curzon asking him to return to London immediately.

But subsequent evidence has shown that Curzon's fate was already sealed. Stanley Baldwin's well-wishers were making Machiavellian moves to see him at No. 10 Downing Street. Less than a week earlier, Baldwin had declared that all he desired in life was to retire to Worcestershire, 'to read the books I want to read, to lead a decent life, and to keep pigs'. If appearances could be realied upon, Baldwin looked eminently suited for the role: with his short, stocky figure, heavy nose and baggy flannels, he seemed far removed from the rapier thrusts of politics. But when the chance for premiership had come, Baldwin had not been reluctant to seize it.

Bonar Law's two Secretaries, Lord Davidson and Colonel Waterhouse, were also busy that weekend playing out fateful deceptions to secure Baldwin's elevation. Even before Bonar Law's return to London, Davidson had tipped off Baldwin about his chief's intention not to name a successor.

On Sunday afternoon when Bonar Law had sent his emissaries

to carry his resignation letter to the King, Colonel Waterhouse slipped in a sealed Memorandum to Stamfordham which heavily tipped the scales against Curzon. The memorandum lies in the Royal Archives at Windsor: Stamfordham's note on the envelope says that Colonel Waterhouse gave him to understand that 'it [the memorandum] practically expressed the views of Bonar Law'. It is now believed that Bonar Law knew nothing about the memorandum.

While claiming to argue the merits of the two candidates, the memorandum shows marked partiality to Baldwin. Denigrating Curzon by saying he 'does not inspire complete confidence in his colleagues, either in judgements or in his ultimate strength of purpose in a crisis', it makes a strong case for Baldwin saying that he symbolizes 'honesty, simplicity and balance'. The memorandum goes on to add 'Lord Curzon is regarded in the public eye as representing that section of privileged Conservatism which...in this democratic age cannot be too assiduously exploited'. As Curzon was passed over on similar arguments, it would not be wrong to deduce that the memorandum struck a death-blow to his hopes.

Colonel Waterhouse had handed over the memorandum to Lord Stamfordham in secret. Bonar Law's son-in-law, Sir Frederick Sykes, who was present at the meeting has said he knew nothing about it. Neither did Bonar Law even though the memorandum was typed in his house and by his Secretary Miss Watson.

Years later, Lord Davidson confessed that he had the Memorandum prepared specifically at Lord Stamfordham's request!

That night Curzon could not bring himself to attend the dinner held for the King by Lord Farquhar at Grosvenor Square. He did not blame the King. But he could not help his feelings. 'It is not for me to explain or find reasons for the King's action. Doubtless he acted for the best. But I think he acted with insufficient consideration for an old public servant.'

Grace went alone to dinner, saying uncharitably, 'George, whose health has always been at the mercy of his emotions, was still too much distressed to dine out. He wanted me to decline as well, but this I said I would not do. . .'

More than the machinations of his detractors, it was this tragic flaw in his character that eventually betrayed Curzon. He was defeated by the punitive anarchy of his uncontrollable will. In India, Curzon struck the death-knell for the Empire because of his overweening and tragic obsession with appearing superior. In the premiership crisis it was not so much his detractors' intrigues that secured Baldwin his success as Curzon's own smug belief that he was above the hurly-burly and the need to stoop to conquer. To understand the enemy within, let us turn to the beginning of Curzon's story.

Chapter 1
The Victorian Scene

George Nathaniel was born on 11 January 1859, the eldest son and heir of Lord Scarsdale of Kedleston Hall in Derbyshire. Curzon never forgot, nor allowed others to forget, the magnificent family manor house and the ancient lineage—the Scarsdales claimed their descent in an unbroken line from the days of William the Conqueror. The family claimed to have come to England from a town in Normandy called Courson, from which they took their name, going on to hold the Kedleston estate for more than 800 years.

Victoria had sat on the throne for twenty-two years at the time of Curzon's birth. She was no longer the frail, tiny girl-queen at whose coronation an elderly peer paying homage had stumbled: forgetting all protocol, she had risen from her throne to help him to his feet. Newly-crowned she had revelled in the pomp and glory of court life, her favourite preoccupation in the evenings being to dance till dawn. The chandelier-lit ballrooms of Buckingham and Windsor were crowded with ladies dripping diamonds and men in full court dress: all this indulgently watched over by her ageing, worldly Prime Minister Lord Melbourne.[1]

Marriage to her cousin changed this. Tall, handsome, earnest Prince Albert from Saxe-Coburg was almost the same age as

Victoria. He was intelligent, sensitive, somewhat dull but from the moment she saw him, she would have no other. An enraptured Victoria gushed over Albert's lovely blue eyes 'which reminded her of an angel', the 'dear lovely face', his 'muscular legs in tight cazimere pantaloons (nothing under them) and high boots'.[2] She had summoned him to her drawing-room at Windsor to propose marriage. He had dutifully kissed her hand, but what his feelings were no one ever was to know.

Now a plump and homely mother of nine children, Victoria was still madly in love with dear Albert. She ruled her Empire with dignity as demanded by her studious, serious husband. Albert shunned the fox-hunting, hard-drinking, philandering members of the aristocracy; subsequently, life in Victoria's court turned sombre and stuffy. Victoria wore high-necked dresses to cover her shoulders in deference to Albert's wishes. She had become very stout and her public expression severe. Her cheeks had pouched, her chin receded though she radiated an awesome dignity.

The Great Exhibition of 1851, the brain-child of Prince Albert, had demonstrated to the British that they held the technological skills of the world. Inspired by the great Frankfurt Fair of the sixteenth century, which had helped to promote local manufacture in his native Germany, Albert had launched the idea of the Exhibition. London's Hyde Park was chosen as the venue and Joseph Paxton commissioned to build a mammoth structure of steel and glass. Neither an engineer nor an architect by training, Paxton created what looked like a vast and elegant summer house; this was scarcely surprising, for he had once been the head gardener to the Duke of Devonshire. It took 2,200 men seven months to complete the Crystal Palace.

In the great hall, which resembled the näve of a church, the apotheosis of British technical skills were put on display— model steam engines, bridges, railways. To the visitors, as they strolled through the exhibition, it suddenly became salient that it was *British* steel mills, blast furnaces, railways and steamships that were opening up the world and bringing peace and prosperity where only ignorance and darkness had once existed. When the Queen and her consort arrived to open the exhibition, they were cheered by a crowd of 700,000 visitors. 'The tremendous cheers, the joy expressed in every face, the immensity of the building, ... and my beloved husband, the author of this "Peace-Festival", which united the industry of all nations of the earth—all this was moving indeed ...', gushed the young Queen.[3]

The exhibits of other nations seemed paltry by contrast, and only heightened the superiority of British manufacture. The weekly takings at the gates were never less than £ 10,000. It seemed as though it were divine will that had entrusted Britain with the duty to civilize the world. The Crystal Palace was a symbol of what British skills could achieve. What better proof could there be of her ascendancy?

By 1859 Victoria had much to feel smug about. The British had vanquished Napoleon and, more recently, in the Crimean War they had trounced the Russian Bear by a striking use of cavalry at Balaklava. The art of the Empire was in place: Landseer had completed his *Monarch of the Glen* and Tennyson his *In Memoriam*. At 40, Victoria could proudly claim that Britannia ruled the waves and the Union Jack flew across a fifth of the world's surface.

Two years earlier the British hold over 'Hindoostan' had been threatened by the outbreak of a sepoys' 'Mutiny'. Once the storm had subsided, the East India Company was dismantled and India brought directly under Victoria's rule. The Governor-General, by now the Queen's sole and direct representative in India, by the royal proclamation of 11 November 1858, was conferred the imposing title of Viceroy.

Some of the tales of horror of the 'Mutiny' had travelled back to the Curzon household in Kedleston. A local soldier, Sir Henry Wilmot,[4] who had fought in it was awarded the Victoria Cross for his bravery. On his return to Derby he lost no time in telling people of what he had seen there. Sir Henry was a frequent visitor to Kedleston Hall and Curzon as a child was to hear first-hand accounts of the wonder of Hindoostan and how the British came near to losing this the brightest jewel in the imperial crown.

London in 1859 was the centre of much that was smart and advanced in European life. Its great opera companies, its symphony and chamber orchestras played every night. Gladstone, Tennyson and Darwin were 50 when Curzon was born. Though Thomas Arnold, the headmaster of Rugby, had died 17 years earlier, Dickens was 47, John Stuart Mill 53 and Palmerston 75. Charles Darwin was about to publish his *Origin of Species*, John Stuart Mill his treatise, *On Liberty*. Wordsworth had died only relatively recently and his poetry was still fresh in everyones' minds.

The fog caused by the industry that contributed to the prosperity of the Empire drifted over the boulevards where the fashionable came to walk. Massive public buildings were adorned with ornamented windows, balconies and coloured doorways. In winter, icy winds and whirling snow swept across the countryside surrounding the city to lash the windows and walls of the city-palaces and to freeze the Thames. Occasionally a brilliant day would break the gloom. The sky would turn a silvery blue and the trees, rooftops and domes would sparkle with bright sunlight.

Summer could be as light as winter was sombre. Windows, opened to catch the river breezes, also brought in the salt air from the Gulf Stream, the aroma of spices and tea, and the sounds of carriage wheels, the factories' blast, the shouts of street vendors, and even the peal of nearby church bells.

Society read Dickens but knew almost nothing of the world of Oliver Twist. They knew little of the appalling poverty of the countryside which had forced thousands to flee to the ugly industrial towns in search of jobs. Many of the poor children still went hungry, few to school. They began working in the factories at the age of ten, in the mines at twelve for six days a week. In the midlands and all around the towns of the north were the signs of that double-edged industrial revolution: the long huddle of textile mills creeping away to the moor's edge and the marching file of tall brick chimneys belching black

smog. The workers crouched hugger-mugger in filthy tenements; so debased were their lives that many thought nothing
of sending out their wives and daughters to eke out a living
from prostitution. They accepted poverty, disease, child
mortality, drunkenness and filth as natural. Many often went to
bed hungry. Workers in the sprawling industrial towns of
the Midlands and the north-west saw little of the growing
prosperity of the time; many were even charged for the use
of lavatories.

The flamboyant Disraeli,[5] with his curly ringlets and purple
velvet waistcoats, who rose to be Prime Minister, knew for
himself the terrible poverty of the agricultural districts and had
been elected by just such a district. In his novel *Sybil or The
Two Nations* he described England as a land of 'two nations',
the classes and the masses. Unlike Victoria's other Prime
Ministers, Robert Peel, Russel and Gladstone,[6] who knew
only the world of the upper classes, Disraeli had intimate
knowledge of both worlds. He had seen the conditions of life
in the teeming industrial towns and had been appalled by them.

In the England of Curzon's childhood, the landowning
families who formed a sort of an exclusive club that ruled Britain,
saw no reason as to why they should give up their primacy.
Their sons were reared in great ancestral piles and studied
at Eton and Oxbridge before taking their seats in Parliament.
The landowners were by and large rich and refined and the
peasantry often ignorant and hungry. The pattern of relationships between the patrician and the plebeian was precisely
etched and was accepted as though it were divinely ordained.
Like most aristocrats Curzon viewed society in these static
terms, terms which remained fixated till the end of his life.

Aloof and governed by strict protocol, the upper classes
followed the rigid schedule of the seasons. Though Curzon's
parents did not follow this merry-go-round, they also scarcely
knew of how the 'other half' lived in the teeming industrial
townships on the eastern outskirts of Derby, even though they
were within driving distance from their manor house. Many of
the famous coalbelts that gave early impetus to the industrial

boom were located there. The building of Midland Railway's Great Locomotive and Coach Works in the mid nineteenth century at Derby had revolutionized the district. But the clank and clamour of machines scarcely penetrated the pastoral serenity of Kedleston Hall, the Scarsdales' estate, only four and a half miles away. All the family knew of the poor was gleaned from the tenants on their estate.

Curzon's own ignorance was rudely betrayed in his first election attempt from South Derby in 1885, which showed he knew little or nothing of the needs and aspirations of the new potters, colliers and factory workers who were added to his constituency with the election Reform Bill. Besides, he felt he had no need to know, smugly believing his ancient lineage and his education entitled him to the seat. Little wonder that they had thrown egg in his face.

But Curzon was not alone in his ignorance. Even Dr Benjamin Jowett, the enlightened Master of Balliol, was pathetically unaware of the masses, confessing in 1861, 'I do little or nothing for the poor, but I have always a very strong feeling that they are not as they ought to be in the richest country the world has ever known. In theory I have a great love for them.'[7] As late as 1923, Sir Harold Macmillan, not without some inverted snobbery, declared:

> I have no practical knowledge of the world in which I was to move. I had never been to Teeside or, even Tyneside . . . I had never seen the great iron-works, shipyards which had been built up on the banks of the rivers or the North of England and of Scotland.[8]

For the vast majority of the upper classes life was a giddy round of pleasure. In London, from May till the end of July, there was a hectic round of balls and dinners. Parties were arranged to witness pigeon-shooting and fashionable flower shows. Between mid-day and two o'clock the park was the most frequented place in London. Here the fashionable world congregated. The ladies wore close-fitting braided habits, the men frock-coats, pearl grey trousers and varnished boots.

Queen Victoria's court might not have been amusing but the

salons of the great political hostesses were. Right up to the Great War, these hostesses played with politics. Lady Palmerston, the Prime Minister's wife, had a salon where you could hope to hear all the gossip.

London was a mad social whirl. Disraeli was reported to have entertained as many as 450 different people to dinner in one week, while another eminent Victorian was known to have once dined out a whole week in advance of his invitations. He discovered his mistake only when a hostess told him that he was expected a week later:[9]

> Sometimes after leaving the dinner table, ladies would not see the gentleman again that evening. Lord John Russel was notorious for this. Free bottles of port (together with the chamber-pots) would be brought out, and every man would drink until most were under the table. It was a regular custom for valets to come in around midnight and carry out their masters. Sometimes everyone was put into the wrong carriages and escorted to the wrong houses.[10]

The Scarsdales came down to London but infrequently. In his childhood Curzon remembers staying at the Burlington Hotel near the Arcade.[11] Later there is mention of a family house in Lower Berkeley Street. While there, somebody in the family always went down with diphtheria or a sore throat, something which was blamed on bad drains. Later Lord Scarsdale took a house at 34 Wimpole Street.

For all its disparities, the tenor of Victorian society remained surprisingly bucolic and calm: both the common man in the street and the rich man in his castle accepted his own lot and that of the other as predestined and normal. Strong Evangelical winds sweeping across Britain developed a keen sense of moral accountancy, promising reward if not in this life then certainly in the next.

Children grew up learning to subordinate instinctual drives to hard work and to doing one's duty, virtues urgently required by a society engaged in rapid industrialization. *Petit bourgeois* rectitude, sacrifice and gloriously earnest humbug were encouraged virtues: qualities which were indispensable in order to hold society together in an era of revolutions. Though workers in the industrial towns saw little of the growing prosperity they were vaguely aware their efforts would one day bear fruit. So they toiled on, their anger and frustration systematically and ingeniously deflected.

Starting in the eighteenth century with William Law, author of *Serious Call*, and continued by school-masters like Thomas Arnold, Evangelicalism really came to full bloom only after Victoria's marriage. Duty and self-denial were the guiding words at Victoria's court and they found echo across the country, even the nobility being gripped by the simple rationale. Every young politician, as he made his way from Eton and Oxford to Parliament, was taught that he was being trained to fulfil his duty to his people and his country.

After the retirement of the suave Lord Melbourne, Evangelicalism touched every prominent Victorian politician save Lord Palmerston.[12] Prime Ministers as varied as Gladstone, Disraeli and Salisbury were both guided by it and exploited it to further their ends; Gladstone with his burning sense of mission being perhaps its finest symbol. Besides, Evangelicalism was proving to be an added boon to the most profitable of all activities—Empire-building. Sacrifices had to be performed without the expectation of gratitude. A similar call had driven Livingstone to explore darkest Africa.

Empire-building was the ultimate white man's burden. Had not the Great Exhibition demonstrated that it was Britain's ships, railroads, ironworks and bridges, the symbols of the technical command of the British that were opening up the dark continents. If their endowment was superior, it was only so so as to enable them to ameliorate the lot of the lesser breeds. Many Victorians actually began to believe that theirs was a moral mission. In 1903 Curzon spoke as one who was personally performing the will of God in India:

> If I thought it were all for nothing that you and I, Englishmen, Scotsmen and Irishmen in this country, were simply writing inscriptions on the sand to be washed out by the next tide; if I felt that we were not working here for the good of India in obedience

to a higher law and to a nobler aim, then I would see the link that holds England and India together severed without a sigh.[13]

Feats of exceptional endeavour demanded exceptional self-denial. Empire-building for many Victorians became a sublimation of the sexual instinct. Libidinal or sexual energy if properly channelized has been known to lead to the highest and most creative human endeavour and to Victorians empire-building was a divine call.[14] Many of the empire-builders married late: Curzon at 36 and Viscount Milner at 67. Kitchener, General Gordon and Cecil Rhodes died bachelors. Sometime the Empire made too arbitrary a demand upon her sons, demanding impossible feats of sublimation. But there was an almost masochistic satisfaction in having done one's duty; a pleasure in self-denial and physical suffering.

In homes and in schools, young boys were nurtured through adolescence to adulthood in a carefully calculated atmosphere where early sex with women was frowned upon. Victorians 'put frilly knickers on piano legs and separated books written by men and women on their shelves'.[15] Attempts were made to steer energy towards greater excellence on the playing field and in the class-room:

> If every value and every force surrounding an adolescent tells him that his bodily affections must at all costs be transformed and sublimated into physical effort, intellectual prowess, competitive zeal and manly friendship, how can he not found empires? If virility is held to be something which is lost, not acquired, in concourse with women, what is left for him but to organise the whole world?[16]

A contempt for early marriage became a part of the public school ideal, empire-builders merely carried their all-male activities to dangerous and far-flung corners of the world. In establishing ascendence over the ignorant, slothful native the pioneers behaved like school prefects carrying out the orders of the headmaster. Lust for women was transformed into lust for conquest. But there were dangers, for like all lust, it brought out the worst in some. Capable of being tyrannical and ferocious, propelled by a messianic zeal, the soldiers of the

Empire could emotionlessly hack their enemies to death without the slightest moral qualms.

Shades of paranoic cruelty were finely balanced by masochistic urges, those of self-flagellation. Like religious fanatics, empire-builders dived into the jaws of danger, ablaze with fervour. At other times relief was found in self-pity and many a manly imperial hero, fearless in battle, dauntless in courage, was seen dissolving in copious tears. At a memorial service for General Gordon outside the ruined palace in Khartoum, national hero Lord Kitchener was so overcome by emotion that he had to order the dismissal of parade. His staff was astonished to see their stern autocratic General shaken with sobs, tears running down his cheeks.[17]

The greatest happiness for the greatest number—the utilitarian creed—by the latter part of the nineteenth century also became another tool in the hands of the empire-builders when seeking a moral justification for their rule in India. Started in the late nineteenth century by Jeremy Bentham, utilitarianism was carried on through to the next century by James Mill, Malthus, Wilberforce, John Stuart Mill, Henry Maine and James Fitzjames Stephen. Both utilitarianism and Evangelicalism had their Indian connections: Charles Grant and John Shore of the Clapham sect having been East India Company servants and Governor-General Cornwallis' advisers on the Permanent Settlement. Grant had advocated aggressive missionary activity in India while both the Mills had worked in East India House, James Mill having written *History of British India*, John Stuart Mill *On Liberty, Utilitarianism* and *Considerations on Representative Government*.

India had provided the laboratory for the experiment; Grant sought salvation for India through Christianity and Macaulay believed that western education alone could transform the natives and lift them from vice bred through ignorance. While James Mill placed his faith on good government and just laws, it was Charles Trevelyan, brother-in-law of Macaulay, who in his pamphlet *Education in India*, put forth the true Liberal attitude to India—which implied that in the eventual analysis altruism was enlightened self-interest.

But the self-assured optimism of the mid-nineteenth century imperialists, which had encouraged western education and forms of government for India, had second thoughts with the 'Mutiny'. Confronted with a rising and articulate English-eduated class in India and Britain's own diminishing superiority in the European arena, both Liberals and utilitarians alike began suddenly to worry about giving liberty to a people not ready for it. Sir James Fitzjames Stephen challenged the intellectual basis of J. S. Mill's doctrine of representative government. Greatest happiness for the greatest number, Stephen said, could best be achieved in India not through Parliamentary democracy but by concentrating power in British hands. He said of the British Government in India:

> It is essentially an absolute government, founded not on consent, but on conquest. It does not represent the native principle of life or of government and it can never do so until it represents a belligerent civilization, and no anomaly can be more striking or so dangerous, as its administration by men, who being at the head of a Government founded upon conquest, implying at every point the superiority of the conquering race, of their ideas, their institutions, their opinions and their principles, and having no justification for its existence except that superiority, shrink from the open, un-compromising straightforward assertion of it, seek to apologise for their own position, and refuse, from whatever cause, to uphold and support it.[18]

With this viewpoint, Fitzjames Stephen makes a complete break from the early Liberalism of Bentinck[19] and Trevelyan which had envisaged not only co-operation between the races but ultimately self-government for India, the country first having to undergo a process of paternalistic rule. Curzon first heard Stephen when in school at Eton and was to be powerfully influenced by him.

Chapter 2

Kedleston Hall

Had not the lusty 2nd Baron Scarsdale decided to make an honest woman out of his beautiful Flemish mistress of sixteen years, Curzon might never have come into his inheritance of Kedleston.[1] The Baron sired six illegitimate children by the lady before settling down to matrimony. It was a great stroke of luck for Curzon that he did. Curzon's grandfather thus became the first legitimate son born to the couple.

It is said that Kedleston gave Curzon his lofty view of life and his sense of destiny. By a strange coincidence, the vice-regal residence in Calcutta, which was completed in 1803, was designed on a model of Kedleston. Curzon first saw Government House as a 28 year-old Tory M P on a visit to India. Upon his appointment as Viceroy Curzon was to remark:

> It is certainly true that it was the fact of that resemblance that first turned my thoughts to the question of the Government of India; and when I left the doors of Government House in Calcutta on the first and only occasion on which I have visited it, in 1887, it made me feel that some day, if fate were propitious and I were held deserving of the task I should like to exchange Kedleston in England for Kedleston in India.[2]

Curzon's father, Alfred Nathaniel, being the younger son of a younger son, had reconciled himself to the rectorship of the family parish at Kedleston. The British laws of primogeniture were such that the bulk of the estate went to the eldest son. The younger sons were forced to seek their fortune elsewhere; the choice being restricted for a scion of the upper classes to the army, navy, politics or the Church. The Empire was also there—offering the more enterprising opportunities of fast money.

Curzon's father chose to become a clergyman. The Scarsdale family had long and continuous connections with the Church.[3] The little medieval chapel at Kedleston crouches behind the manor house, the church graveyard being full of memorials to the family and its entourage. Five-hundred-odd parishioners were tenants of Kedleston. Alfred Nathaniel was happy with the gentle tenor of a parsonage. Within a year of his rectorship in July 1856 he married the 19 year-old Blanche Senhouse. Her father, Joseph Pocklington Senhouse, of Netherhall belonged to the landed gentry and not the peerage.[4]

The marriage customs among the British landed classes were then such that the bride was expected to bring as dowry a sum of cash called a 'portion'. In return the head of the household was expected to guarantee the bride an annuity called a 'jointure' if she survived her husband as a widow. The result was that unless the younger son had made his own fortune, he was rarely able to attract a wealthy bride. But that same year, by several quirks of fortune, Alfred Nathaniel suddenly found himself catapulted into becoming the 4th Baron Scarsdale of Kedleston Hall.

His ancestor, the bohemian 2nd Baron had created untold complications for the Scarsdale family. On coming of age, he had married and produced a son, Nathaniel, who succeeded him as the 3rd Baron upon his death in 1837. The 3rd Baron, however, remained a bachelor and in the normal course of affairs the title would have passed to the next brother. But here arose the problem created by his father.

After the death of his first wife, Sophia Noel in 1782, the 2nd Baron had taken to heavy gambling. Finding himself deeply in debt he ran away to Europe to escape the clutches of his creditors where he fell in love with the Flemish Felicite Anne and, fathered six illegitimate children by her. He finally married her in 1798, the ceremony taking place in a town called Altona near Hamburg. Four more children were born to Lord and Lady Scarsdale after the marriage, who being born in wedlock were not debarred from succession. Curzon's grandfather was thus the first legitimate son born to the second Lady Scarsdale, though he was actually her sixth son.

Still, the course of succession was hardly smooth. Curzon's own father, Alfred Nathaniel, was merely a younger son. His older brother George Nathaniel was unmarried though suffering from pangs of unrequited love. He was in the habit of wandering around London's Rotten Row by Hyde Park, the place where the fashionable world of London assembled and where love affairs flourished accompanied by the ceaseless rustle of delicious gossip. George Nathaniel's chief purpose behind frequenting the Row was to catch a glimpse of his former lady-love who had spurned him for a wealthier suitor. One day, he thought he was in luck, for just as he was leaving the Row, the fickle lady passed by with her mother in their carriage and condescended to toss him a nod. Breathless with excitement, he tried to reciprocate, but the movement proved too much for his horse who reared up, fatally dashing his mount against the curb. Thus, the premature and unexpected death of this uncle clinched Curzon's inheritance.

Before Curzon's father could become the new 4th Baron, there were further hurdles to cross. Two illegitimate sons of the 2nd Baron were still living in 1856. If it could be shown that the 2nd Baron's marriage in 1798 was not valid in English law, the estates would pass by a family settlement of 1816 to the 2nd Baron's illegitimate issue, in order of seniority. The two living illegitimate uncles who stood to benefit were Edward Curzon, a retired admiral and Frederic, Rector of Mickleover. But in 1857 Curzon's father was able to prove in the English courts the validity of his grandparents' marriage in 1798, which made him at 25 the rightful heir to the peerage and the estates.

The eighteenth century has been widely accepted as a sexually promiscuous age, especially among the aristocracy. The earier collapse of moral puritanism as a dominant influence in society after 1660 led to the general secularization of society. Prominent among its features was the release of the libido from the former restraints of Puritan Christianity,[5] one of the marked changes being a rapid increase in extramarital sex among the court aristocracy, which soon spilled over to the rest of the élite.[6] Dramatist Richard Sheridan said, 'In Oliver Cromwell's time they were all precise canting creatures. And no sooner did Charles II come over than they turned gay rakes and libertines.'[7]

Rakes were tolerated by society so long as they provided for their illegitimate brood. In 1790, for instance, there were a whole collection of oddly assorted children being brought up at Devonshire House and Chatsworth: among them were the children of the 5th Duke of Devonshire and the Duchess Georgiana, two of the Duke and Lady Elizabeth Foster, the Duchess's most initimate friend and lifelong companion, while one child of the Duke and Charlotte Spencer and one of the Duchess and Lord Grey were brought up elsewhere.[8]

The pornography and erotic poetry of Marquis de Sade were then popular in France. In England, Sir Francis Dashwood had set up the Hell-Fire Club. Its counterpart, the Society of Beggars Benison, was set up in Scotland in 1732; its mottoes being 'Be Faithful and Multiply' and 'Lose no Opportunity'.[9] The 2nd Baron had been a member of the Beggars Benison and Curzon nicknamed him thus. In 1921, writing to Grace, Curzon recounted:

> I have come across all the papers about the marriage of Beggars Benison with Felicite Wattines. They had 5 sons and 1 daughter before they were married in 1798 when living in Altona just outside Hamburg. From the papers it appears that he thought that under the law of Hamburg, then a Free State, this would legitimise the children already born. He was finally married in the House of the Rector at Hamburg: and then arose the question—was the record of the marriage without banns and without an Entry in the Register of any Church a valid marriage? This was fought out in

the law courts and in the affirmative.

By the time that the 3rd Lord who was the eldest son of Beggars Benison and his 1st wife died in 1856 and my father succeeded— only 2 of the illegits were living and they, being very fond of my father who gave them both a handsome annuity, decided it was no good fighting the case which they would certainly have lost.[10]

The 4th Baron, Revd Alfred Curzon, was born on 12 July 1831. He was schooled at Rugby, made famous by Dr Thomas Arnold, who maintained that the school's primary function was to first improve character and then the intellect. When it came to a choice between scruples and science, Dr Arnold had favoured the former. Alfred Curzon later proceeded to Merton College, Oxford, where he took his BA in 1852 and MA thirteen years later. In 1854 he received the Deacon's Orders and was made a priest in 1855. A year later, he became Lord Scarsdale, with the death of his uncle. He died at the ripe age of 85 in 1916, leaving four sons and six daughters. Curzon was 57 when he inherited his father's title and Kedleston Hall.

Though he studied at Rugby, Lord Scarsdale sent his sons to Eton. As Curzon had only daughters, after his death in 1925 Alfred's son, Richard, became the 2nd Viscount Scarsdale.[11] Till as late as 1923, two years before his death, Curzon did not give up hopes of producing a son and heir. His second wife Grace co-operated by going to the German spa, Langenschvalbach, for treatment. From there she wrote, 'I was thoroughly examined by the doctor . . . He is *delighted* with me. He says it will not be my fault if all our hopes are not realised.'[12]

A year later Grace wrote, 'I am feeling *so* well and still full of hope—and so much love waiting for the future Earl of Kedleston.'[13] They were to be disappointed once again. No child was born from the marriage.

In the meantime, Curzon kept up a vigilant watch on the newborns in the rest of the Scarsdale family. As late as December 1923, he was reporting with malicious relief the arrival of a female child among his relatives:

Frank rang up yesterday to say that Mrs Dick had produced a daughter. The dates justify all our conjectures: for he was only married on April 14: so that not 8 months have elapsed: Looking up my papers I see that it was on March 7 exactly nine months before, that he wanted to be married at Rome. Therefore the whole thing is clear. The woman came down to meet him at Rome and either said she was already in the family way or made him sleep with her there, and then extracted from him a promise of immediate marriage ... I am wondering whether they will dare put the birth into *The Times*. (He dare not!) anyhow it is a girl, which is a good thing.[14]

Curzon's father, on becoming the 4th Baron, did not divest himself of the Holy Orders. He combined his new responsibilities as a landowner with his vocation. His children, however, were to complain that while he would spend money on repairing a labourer's cottage, he would turn a jaundiced eye to any suggestion for renovating Kedleston. Curzon's first wife Mary wrote, 'He looks at the dilapidated rooms with grim satisfaction.'[15]

Kedleston with its 10,000 acres was small compared with some of the other estates in the country. A survey of 1870 shows that 44 individual landowners had more than 100,000 acres each, many of them living in extraordinary splendour. Lord Egremont's stables boasted 300 horses. At Inveraray Castle, the Duke of Argyll's guests were summoned to meals to the call of bagpipers. The opulent style of living of the Duke of Sutherland so alarmed the visiting Shah of Persia that he cautioned the Prince of Wales: 'Too grand for a subject. You'll have to have his head off when you come to the throne.'[16]

Lord Scarsdale shared the family pride for Kedleston and it is a mansion worthy of it. In its vast marble reception hall, twenty soaring columns of pure alabaster rise high to a gilded ceiling. As in the atriums of imperial Rome, ornate alcoves alternate between the pillars. But instead of illustrious ancestors, statues of classical gods adorn them at Kedleston. The real masterpiece, however, is the beautiful circular salon to which the Marble Hall leads the way. The walls suddenly soar upwards to a total of sixty-two feet and gently curve into a golden dome.

Huge galleries fan out of either side of the main building to two smaller supporting wings, leading to a state boudoir, a drawing room, dining room, bed-chambers and the library. Tall windows gaze down on a vast parkland enclosed by massive wrought-iron gates, emblazoned with the family crest in gold, and the gravelled road which winds its sedate way from Kedleston between a phalanx of stately oaks.

The earlier house had been a three-storeyed manor in red brick. But with new money from investments in the East India Company and the American colonies, the 1st Baron Scarsdale felt the existing house to be too modest a dwelling. Besides, being an avid collector of paintings, he wanted a better setting for his portrait gallery; the Van Dycks and the Reynolds are still on display.

Sir Nathaniel decided to demolish the old house and build a new. Architects raised a problem. To build the sort of dwelling desired, vast open spaces were required. It would entail shifting the village of Kedleston. But this was not a detail to worry the 1st Baron. He promptly approached his friend and monarch, George III and the King obliged in a fittingly regal gesture. By an Act of Parliament the public highway was diverted and the entire village shifted to a mile away from the required site. Thus began the building of Kedleston.[17] Curzon's monumental style can perhaps be traced back to this grandiloquent ancestor.

Three architects moulded the house, which was commissioned by the 1st Baron Scarsdale. Brettingham and Paine envisaged the main plan of a central pile with its extended wings. A third was brought in, initially to design the carpets and ceilings for the interior. Robert Adam, youthful, inexperienced but bright, was at the time nowhere near making his name as one of England's foremost architects. But the other two architects, being more famous, got busy with different commitments and left Kedleston to Adam. It was Adam's genius that transformed it from a mere wealthy nobleman's residence to one of the prized stately homes of England.[18]

Besides designing the building, Adam exuberantly fashioned gilt sofas with dolphin feet and cast-iron altars in the boudoir alcoves to ingeniously conceal a heating system. He decorated mirrors with exotic gold palm-leaf motifs and in the bed-chambers carved bed posts to resemble palm-trunks. He also turned his fantasy to the parkland, landscaping it with a natural running brook and spanning it with a graceful three-arched bridge.

Such grandeur was bound to arouse jealousy and Horace Walpole after a visit to Kedleston in 1768 said, 'A fine park with old timber, beautiful gateway with lovely iron gate by Adam, a vast house with four wings, of which two only yet built, and magnificently finished and furnished, all designed by Adam in the best of taste but too expensive for his estate.'[19]

True, the Scarsdales had never been very wealthy nor were they particularly distinguished. Curzon said, 'My ancestors have held Kedleston for 900 years, father and son, but none of them ever distinguished himself. They were just ordinary country gentlemen...M. R. Sheriffs, and so on.'[20]

The estate was heavily mortgaged when Curzon's father inherited it. But with his economy Lord Scarsdale was able to leave a gross figure of £ 454,694 in his will.[21]

Irene, his grandchild and Curzon's eldest daughter, came out warmly in the old man's defence. 'My grandfather put every penny of his income to its best use. Having to provide for his vast family he could only give his eldest son a modest allowance, which necessitated a strict supervision of expenditure.'[22]

Few toys were allowed in the house of Kedleston. One of Curzon's sisters told his daughter how the Scarsdale children taught themselves the alphabet by laying out their scarves and coats as letters on the floor of the cold and drafty great hall.[23] A sign in the kitchen said, 'Waste not, want not', but it would hardly be true to say the house was run on the lines of a country vicarage as has been claimed.[24] The family generally lived quite comfortably, dining off silver plate, there being enough for seventy-two people.[25] Waiting upon them was a hierarchy of servants befitting a nobleman's residence. Curzon's first wife, Mary, remembered counting twenty-three servants when she came as a bride to Kedleston in 1895.[26] They, however, were probably only the 'upper servants'. The 'lower servants' who did the menial and lowly jobs, generally ran to dozens, but were rarely seen or heard in the great houses of England. Being a kindly master, when Curzon came of age Lord Scarsdale thoughtfully gave 'a servants' dance'.[27]

During the holidays the family took regular vacations either to the seaside or to London, where they stayed at the fashionable Burlington Hotel off Piccadilly. The trips to London were in the nature of family outings. Lord Scarsdale had little interest in taking his seat in Parliament or in the high society of London.

Life at Queen Victoria's court had virtually come to a standstill with Albert's death in 1861: the Prince Consort having died from a chill caught on a trip to Cambridge where he had gone to reprimand his eldest son for getting entangled with an actress of easy virtue. Victoria never forgave her son, and she went into mourning for Albert, which was to last till her death in 1901. Victoria became a recluse, spending much of her time in morbid preoccupation with death, putting up memorials to her husband. Every day she had Albert's clothes, hot water, and fresh towels laid out in his room at Windsor and went to bed clutching his old night-shirt.[28]

The royal mourning, however, did not unduly disturb the tenor of Scarsdale lives. Lord Scarsdale was happiest being at Kedleston, firmly believing that the duty of a landowner was to stay on his land and among his tenants. Having not given up his Holy Orders, he preferred to call himself the Revd Lord Scarsdale.

With a cleric as father, and without the godsend of the family title and property, Curzon might have grown up at the rectory, which along with the cottages of the tenants, the malt house, the forge and the mill, had been moved a mile away in order to clear the parkland for the building of the new Kedleston Hall. Curzon prized the accidents of fate which brought him into his inheritance and spent much of his life trying to maintain an affected aristocratic style which he felt behoving of his status; at times he seemed to turn this into parody.

Chapter 3

Imagined Tormentors

Over the years an impression has built up that Curzon's parents were callously indifferent to him: that his childhood was blighted by an uncaring mother, a coldly stern father and a savage governess. Such was the malignancy of these infantile influences that, it is believed, they darkened forever his subsequent growth and behaviour. His mother, Lady Blanche Scarsdale has been particularly singled out and castigated for rejection of her son. One biographer says of Blanche:

> When she was shown her son and heir for the first time, she . . . looked at her top-heavy offspring with the cold surprise that was to be the quality of her attitude towards her son for the rest of her life. She handed him back to his wet-nurse and asked that he should be brought back for further inspection the following morning. That was to be the pattern in years to come . . . As he grew older, and began to yearn for the affection which his nature increasingly needed, George Nathaniel Curzon became slowly aware that beneath the surface his mother's attitude towards him was so detached as to be almost indifferent.[1]

Another is less harsh but he too does not absolve Blanche of her neglect of her son:

> Lady Scarsdale . . . kept a diary during succeeding years. It contains details of family comings and goings, the weather, texts of

sermons, the clothes she wore each day and successive winners of the Derby. But there is nothing of her intimate thoughts, not a word to suggest that her children ever caused her anxiety or brought her pride. She remains an elusive figure.[2]

What is more amazing is that Curzon himself contributed to this impression, when he said, 'I suppose no children so well born or so well placed ever cried so much or so justly.'[3] He was to imply that he was unfairly treated, that adult tormentors had conspired to crush his ego and break his spirit. Of his father Lord Scarsdale, Curzon implied that he did not share in his son's ambitions and took little interest in his career. To friend Lord Riddell, Curzon said, 'My old father...had no sympathy with my aspirations. I don't believe that my father ever read one of my books or speeches, and he took very little interest in any success I achieved.'[4] Curzon liked to call himself a self-made man, implying that he had achieved his success despite the handicaps put in his way.

In the pencilled notes written for his biographer Curzon adds to the myth, perhaps subconsciously, when he talks about the indignities heaped upon the Scarsdale children in the nursery by a brutal and vindictive governess. Talking about her, he says, 'She persecuted us and beat us in the most cruel way and established over us a system of terrorism so complete that not one of us ever mustered up the courage to walk upstairs and tell our father or mother,'[5]—leading the reader to believe, however inadvertently, that the parents did not care to know what was happening to their children in their own home and therefore must take equal blame.

If such was truly the case, Curzon's parents have much to answer for. Not only would they deserve censure for parental neglect but also for having given blanket control of their child to a governess without making proper inquiries about her nature and temperament. History would hold them responsible for Curzon's increasing authoritarianism in adult life, tracing it to the ego shattering excercises of his childhood. The Viceroy's compulsive need to crush all opposition to his rule in India could only have stemmed from those infantile tussles when the authority in command snuffed out early rebellion. Curzon's obsession to masquerade as a superior person would be understood as a bid to capture the attention which had never been forthcoming in his vulnerable and stormy years of growing up. But evidence has shown that the reality was astonishingly different. Far from being harsh, let alone indifferent, the Scarsdale parents doted on their eldest son. He was their pride and joy and from the moment of his birth, they lavished upon him a degree of care and attention that was nothing short of extraordinary.

Among the Curzon papers at India Office Library there is a hand-written note by Blanche Scarsdale which makes nonsense of any theory that she did not care for her eldest son. A curious, moving document it is an hour-by-hour account of Curzon's birth and the couple of days thereafter.[6] Blanche's jottings record the castor oil taken to induce pain, the birth-pangs as they came, the consternation at the baby's arrival before the doctor or the midwife could reach her, and her young husband's concern that his two-year-old daughter should not be forgotten in the excitement.

Neither did his mother show 'cold surprise' nor hand him to a wet-nurse. Blanche Scarsdale breast-fed her son. 'No fullness at all in my breasts yet,' she complains at first. 'I only feel the drawing throbbing pain when baby sucks but this leaves off by Friday—and I begin to wear my metal nipple shields.'[7]

What makes Blanche's action particularly significant is because Victorian upper-class wives rarely breast-fed their children and husbands generally discouraged them from doing so, fearing that it might spoil their figures. Writing in 1792, Mary Wollstonecraft said, 'There are many husbands so devoid of sense and paternal affection that, during the first effervescence of voluptuous fondness, they refuse to let their wives suckle their children.'[8]

The campaign against wet-nurses had begun way back in the early seventeenth century when influential Puritan writers on household management, like Perkins, Cleaves and Dod, pleaded with mothers to feed their children, arguing that nature had provided breasts for feeding. In the early eighteenth century, their appeals were powerfully reinforced by fierce attacks in newspapers on the practice of handing children over to slovenly and dirty wet-nurses. In 1748, Dr William Cadogan published his widely read *Essay Upon Nursing and the Management of Children.* He demonstrated that 90 per cent of the children who died were reared on pap and fed by lazy wet-nurses with a poor or contaminated milk supply.[9] Instances were cited—the Duchess of Devonshire feeding her eldest son for nine months. Nevertheless, the institution of wet-nursing not merely survived but flourished. At the turn of the century, Curzon's wife Mary was to take out a wet-nurse to India for her second daughter. 'Cynthia is so beautiful and angelic and the wet-nurse flourished, so all is well,' Mary reported in her journal.[10]

George was born at eleven in the morning before the doctor could reach there, but Blanche's mother was also present at the childbirth. Contrary to some accounts, Blanche had a smooth and easy delivery.[11] 'I was very well indeed Monday night,' she reports of the eve of Curzon's birth. She took castor oil and by Tuesday morning she felt labour pains: 'positive labour, so I tell [Alfred] and he send a note to Dr Gisb by groom telling him to meet him in Derby by 12.'[12] Another message was sent to the midwife Mrs Blundstone. But suddenly the pains increased, '... every 5 mins almost mother [Blanche] felt sure it was coming on quickly and fresh messages sent off to bring Gisb *at once*—but all over before either he or Blumy come.' At ten minutes to eleven the water bag burst and Blanche's mother and husband wheeled her into her room. Soon the baby was born. 'A little boy this time,' she happily declared.[13]

Within half an hour of baby's delivery, the young father was persuaded by Blanche to go to a meeting of the Diocesan Church at Derby, but on his way back he thoughtfully stopped to buy a doll for his elder child, two-year-old Sophy, so that she did not feel neglected in the excitement of the arrival of a brother. Blanche records with shy pride that at Kedleston, 'Bells rang on...our little son and heir's birth.' She notices,

'His face is very red and little head so hollow on each side, mostly on one—so very queer—plenty of space for filling out into a big pate.' The baby is brought every few hours for his feed, 'smiling so sweet and so dear . . .' in his Blackburne lace cap which just fits him'. Once when the baby cries a great deal because of a stomach upset, she blames herself saying, 'afraid it is because of my plum eating last night'.[14]

George was a fair baby with curly thick brown hair, clear brown eyes and an abnormally large head. In later life Curzon would recall that because of the large size of his head, 'It was considered unsafe to leave me anywhere near the top of a staircase because on one occasion I was overbalanced by the momentum of that article and rolled down from top to bottom'.[15] Nevertheless, he must have been an exceptionally handsome child. As he grew older his mother, fancying him as another little Lord Fauntleroy, took him on regular trips to London to have him fitted in velvet suits at the fashionable Swears and Wells, tailors and outfitters.[16] She also had his hair grown long. Curzon wrote in his memoirs, 'My hair was kept in corkscrew ringlets brushed round the finger with a hair brush, and falling upon the shoulders. I remember I cried very bitterly when it was cut off.'[17] His mother preserved these locks as parts of her collection.[18]

She also preserved a miniature portrait of her son in long corkscrew curls wearing a white muslin dress[19] and a sealskin waistcoat. It was later thoughtfully handed over by Lord Scarsdale to Curzon's first wife Mary.[20] Notable among the collection is a letter written by one year-old baby Curzon to his aunt Mary Senhouse upon her engagement![21] His doting mother tucked it away among her prized possessions. In the sixteen years after Curzon's birth Blanche was to have nine more children, of whom one died in infancy.

Biographers have not been very kind to Lord Scarsdale either. One says,

> For him the three great rewards of his life were, in order of import-
> ance, his peerage and the ancestry behind it, the pulpit of Kedleston
> Church from which he preached his weekly sermon, and his

wife. He loved them all in diminishing stages, and these, together with his relentless pursuit of the rich game in his parklands, did not leave him much time for his eldest son or the children who followed.[22]

Curzon's first wife Mary has added to the general impression, calling her father-in-law, 'the most tyrannical old man I have ever seen, besides being the most eccentric.' She elaborated,

He could not tolerate having any friend of his son stay in the house. He could not endure a minute's unpunctuality. Nobody dare lift a spoon without his permission. He viewed with displeasure any scheme for modernizing the house, unwilling to admit that it would one day pass to his son.

According to Mary, he scolded his daughters for not being married but then never allowed a young man to enter the house. 'He is an old despot of the thirteenth century,' she declared.[23]

This was hardly true. Lord Scarsdale did not discourage his son's friends from coming to stay at Kedleston, the most frequent among them being the controversial Eton tutor Oscar Browning[24] and St. John Brodrick.[25] Curzon wrote to Browning, 'We have some people in the house for balls on Tuesday and Wednesday' and added an invitation for him to join them.[26]

Nor were the Scarsdale daughters neglected. Records show 'coming out' balls organized for them. 'Last night I was at a most enjoyable dance in Nottinghamshire,' Curzon wrote, 'it was my sister's debut.'[27] Besides, when Curzon first brought his bride home to Kedleston in 1895, Lord Scarsdale welcomed her in the lavish manner befitting a future lady of the manor.

In spite of all the stories of Revd Lord Scarsdale's legendary frugalities, his children lived well,[28] the family taking vacations in London, where the children were taken to the tailors to be fitted for new clothes. There were also trips to the pantomimes and to Madame Tussaud's.[29]

In Curzon's childhood Kedleston may have been little of the lavish entertaining customarily carried out in the great homes of the nobility, but friends and relations were constantly coming to stay. The earliest diary kept by the seven year-old Curzon has been marked in pencil with the birthdays and wedding days of parents, grandparents, uncles and aunts. Blanche's relations came often and were made very welcome. Lord Scarsdale would sometimes travel to London to bring grandfather Senhouse with him. 'G.papa gave me a pretty little glass stand with Decanter, & 6 glasses, for liquor.' Three days later when the old man left, Curzon wrote, 'We were very sorry'.[30] But the Scarsdale grandmama was not so popular. There is an exclamation mark behind the entry recording her return from Aunt Sophy's 'home'![31]

Every year the family took a holiday at the seaside.[32] It was an extravagant migration. Recording one of such occasion, Alfred Curzon a younger brother, noted in his diary: 'The yellow piano, perambulator, Papa's writing table and different boxes etc. come later by a luggage train. Such work to get piano in school room and up the stairs.'[33]

Photographs show Lord Scarsdale to be a thin, tall, taciturn-looking man with a somewhat severe face, framed by side-whiskers. These, however, were not as big as those favoured in the pictures of Crimean War heroes and which were the fashion of the day.[34] He is seen generally wearing a sober black frock-coat or tail coat, tubular trousers and a top hat. A somewhat gruff, blunt man, he disliked pomp and vanity and exhorted his children to be simple and God-fearing.

Like many a nobleman of his times, Lord Scarsdale had his share of eccentricities. Mary was horrified on her first visit to Kedleston to find him making fiendish grimaces in the mirror.[35] In the midst of winter he slept with no blankets even though he kept eighteen thermometers in his sitting room. Beneath these quirks he was an extremely shy man who was unable to easily articulate his feelings. It is perhaps due to this inability to be demonstrative that Lord Scarsdale has been condemned as a cold and uncaring parent.

To understand him better it is necessary to see him in the context of the traditional Victorian child–parent relationship: the Evangelical bourgeois parent of late eighteenth and early nineteenth centuries being but an extension of the Puritan bourgeois parent of the seventeenth, faithfully continuing the tradition of a caring but authoritarian discipline.[36] Both supported the idea of original sin: children were born with evil and it was the duty of parents to crush it out of the system with strict repression. The father, as the head of the household, was to do his task dutifully, though the internalized Victorian guilt-complexes and soul-searching led to curious results.

Lord Scarsdale demonstrated some of that quality of extreme reserve, bordering on coldness, which the traditional upper-class Victorian parent seemed to have reserved for his children. He could be stern, the dangers of vanity being a pet obsession. When Curzon received the Prince Consort French Prize, Lord Scarsdale had said,

> Your success has given us all the greatest pleasure. You deserve credit I am sure will receive it, but do not, dear Boy, be unduly puffed up: at your comparatively early victories—your talents are given to you by the Almighty God and I fervently pray you will ever use them rightly, and be a comfort and blessing to your parents.[37]

The pecuniary instinct, never far below the surface, asserted itself, making him add, 'let us have all particulars, number of competitions, value of Prizes.'[38] Lord Scarsdale also had a fetish for accounts and details, a trait he was to bequeath to his son. After his death in 1916, Curzon said while making an inventory of Kedleston and his father's sixty year dominance, 'he kept every bill even for 1/6 or 2/6. There is every detail about the great estate and thousands of other things. There are all the school room and school books of every member of the Curzon family for a century.'[39]

Otherwise, Lord Scarsdale deviated vastly from the familiar Victorian mould, his relationship with his eldest son being characterized by rare tenderness and intimacy. He was extremely proud of him and in 1887 when Curzon, on his first grand tour abroad, wrote an account of the British mission in the Far East, Lord Scarsdale sent the letter to Prime Minister Salisbury. Among his papers there is an official acknowledgement.[40]

It has been claimed that Curzon held his father in awe and

fear and was not able to enjoy a normal father–son relationship. The instance has been cited of how at the age of 36 and as an Under-Secretary of Government Curzon was nervous about breaking the news of his planned nuptials to his father.[41] His father, on the other hand, had received the news very well, saying, 'So long as you love her and she loves you—that is all. You are not likely to make a mistake at your age, and she is old enough to know her own mind.'[42] An overjoyed Curzon rushed off and wrote to Mary as if a great load had been lifted off his mind. 'I had to make none of the apologies or explanations or defences that you imagined,' he was supposed to have said.[43]

Curzon's diffidence in this particular case, however, was not wholly unwarranted. In an age when parental consent was a vital prop to matrimony, he was marrying a foreigner and one with a Jewish-sounding name. Besides, so strong was the bond between father and son, that Curzon felt he had to have his father's approval for all his decisions. Age-wise, as his father had pointed out, Curzon was grown up and a free citizen. Nevertheless, he cared sufficiently to want to have his father's approval. He had shown his father Mary's photograph and was delighted when the old man said it was all 'right and proper'.[44]

Thus, what emerges is that in reality Curzon had a very strong sustained relationship with both his parents. In the earlier years it was his mother who had aroused his adoration; after her death it was his father whose recognition and affection he craved for. Freud maintains that one's early sexual impulses are directed towards the parent of the opposite sex. Boys attach their first affection to their mother and develop a rivalry with their father, who is the possessor, as it were, of the mother—a desire which is frustrated by biology and incest taboos. The consequence of this is a strange sense of guilt.[45] To say that Curzon as a child showed practically no hostility towards his father is not to say that his mother was not the object of a deep sensual attachment for him. As a child, his letter from his first school at Wixenford breathes an intimacy with the mother that excludes the father. The mother is used as a channel for requests and she, by tacit acceptance of this role, indicates to him that the two are part of a secret pact. He is playful and daring with her, a spirit that does not come out in the letters to his father. In a moment of exuberance he once naughtily rounds up a missive with:

> Believe me your loving son,
> Georgie Porgy put in a pie,
> He kissed the girls and made them cry.[46]

Curzon may not have taken such liberties with his father. But he behaves like one who is confident that his father shares his mother's love and admiration for him. It appears that Curzon held his father in great esteem and strove hard to be the comfort and blessing that Lord Scarsdale hoped he would be. Though his father might have caused him discomfort by constantly warning him about not getting a swollen head, Curzon nevertheless unhesitatingly turned to him when in need or in trouble and the old man did not seem to have failed him. At Eton, when master Oscar Browning was accused by the Headmaster of imposing his 'irrepressible attention' on Curzon, it was the stern Lord Scarsdale who came to the rescue, declining to pay attention to the scandal.[47] When Curzon got his first cabinet appointment he immediately sent off the Prime Minister's letter to his father, 'knowing how keen an interest greater I truly think than my own you have taken in my promotion.'[48] On another occasion Lord Scarsdale's congratulations on Curzon's appointment to the viceroyalty sent Curzon into a flood of tears; Curzon's reaction almost conveying the impression that the whole effort of achieving the viceroyalty had been geared towards receiving such a recognition. Mary testifies that her husband's outburst of emotion upon receiving his father's congratulations was 'the happy sobs of a son who has been recognised at last by his own.'[49] It was as though Curzon went through his life looking over his shoulder to reassure himself of parental love and admiration.

Curzon continued a steady contact with his father after his marriage even though he could not have been unaware of Mary's antagonism. Lord Scarsdale was invited to stay at

Curzon's home in the country, The Priory.[50] When Curzon
rented Inverlocky Castle in Scotland in the autumn of 1896, a
year and a half after his marriage, he invited his father along even
though he knew that Mary did not much care for the arrangement.[51] He opened his heart to his father about his most private
fears and sorrows. When Mary gave birth to a third daughter, he
told his father, 'It is a blow her child being a girl. . . But she
will feel it more than I since in these matters a man philosophises
whereas a woman cannot. She seems to have got through
splendidly which is the main thing.'[52]

The first mention of his governess, Miss Paraman, appears in
1866. On 21 April Curzon jots in his diary: 'Miss Paraman took
Sophy and me to see Miss Wilson. Mama took Affy. and Blanche
and Welch to Derby.'[53] Miss Paraman, in Curzon's words had,
'grey eyes, thin hair and a large thin mouth'.[54] Curzon's daughter
Irene adds to the description by telling us that she was in the habit
of wearing 'long, voluminous skirts edged with braid'.[55]

Pictures show Curzon about the age of six as a bright, sensitive
and lively child. He probably had turned to his new governess,
as was natural, looking for affection and attention. Governesses
were an integral part of British upper-class life, tradition
relegating the mother's job to them. The children would
be totally in her charge and appeared before their parents
briefly at breakfast, at lunch and shortly before bedtime.
Winston Churchill was to write glowingly about these women
who brought up their charges with devotion and care: 'It is a
strange thing, the love of these women. Perhaps it is the only
disinterested affection in the world. The mother loves her
child; that is maternal nature . . . but the love of a foster-mother
for her charge appears absolutely irrational.'[56] But then Churchill
had been blessed with the plump and motherly Mrs Everest
who seemed to have belonged to a totally different category
from Miss Paraman. Churchill said of Mrs Everest, 'My
nurse was my confidante. Mrs Everest it was who looked after
me and tended my wants. It was to her I poured out my many
troubles.'[57] Because of Mrs Everest's reassuring presence, it
did not matter that his own brilliant, beautiful, socialite

mother Lady Randolph Churchill should appear infrequently. He was to say lovingly of his own mother, 'She shone for me like the evening star. I loved her dearly—but at a distance.'[58] Miss Paraman, however, as Curzon has recalled:

> ...shut us up in darkness practised upon us every kind of petty persecution, wounded our pride by dressing us (me in particular) in red shining calico petticoats (I was obliged to make my own) with immense conical cap on our heads, round which, as well as on our breasts and back were sewn strips of paper bearing in enormous characters written by ourselves the words Liar, Sneak, Coward, Lubber and the like. In this guise she compelled us to go into the pleasure ground and show ourselves to the gardeners. Our pride was much too deeply hurt.[59]

Victorian children were not unfamiliar with this humiliation of being made to parade with placards stitched to their backs as Kipling testifies in his story 'Baa Baa Black Sheep',[60] but Miss Paraman made beating a normal part of childhood experience. Curzon has said,

> She spanked us with the sole of her slipper on the bare back, beat us with her brushes, tied us up for long hours in chairs in uncomfortable positions with our hands holding a pole or blackboard behind our backs.[61]

What seems even more shocking is that she punished them for sins they had not committed. To break the innate pride of her charges, she used humiliation as her controlling device:

> She made us trundle our hoops, all alone, up and down a place in the grounds near the hermitage where were tall black fir trees and a general air of gloom of which we were intensely afraid. She forced us to confess to lies we had never told, to sins which we had never committed and then punished us savagely as being self-condemned. For weeks we were not allowed to speak to each other or a living soul. At meals she took all the dainties herself and gave us nothing but tapioca and rice.[62]

But what perhaps appears to us as most tragically shocking is the Scarsdale parents' total oblivion to the indignities heaped upon their children and the latters' inability to confide in them.

'She persecuted and beat us in the most cruel way and established over us a system of terrorism so complete that not one of us ever mustered up the courage to walk upstairs and tell our father or mother,' says an agonized Curzon.[63]

Horrifying as the catalogue of punishments sounds, taken in the Victorian context it appears plausible. Biographies of that period are full of accounts of flogging and repression in school rooms. Victorian upper-class children are known to have lived in stuffy nursery quarters separate from the rest of the house. It is generally maintained that, '...the Victorian nursery as it finally crystallized' was 'an austere area, furnished with furniture not needed elsewhere. Austere and remote—situated at the top or in far-flung parts of the house, among servants. When there was a fire it was the children or servants who got burnt.'[64]

Governesses were in a position to play havoc with their charges. They had keys to the medicine cupboards and could administer dangerous opiates to drug the child.

> Wine was frequently prescribed by doctors; but it was left to the nurse to decide when the time had come to force down a reluctant throat a pill or powder, a dose of brimstone and treacle, caster oil, liquorice or a spoonful of the notorious Godfrey's Cordial, that mixture of laudanum and syrup which, together with other similar products, could be purchased at any chemist's shop and which reduced children to varying states of stupefaction for hours on end.[65]

Victorian children seem to have had to take an amazing amount of battering of both mind and body as part of the process of growing up. Social historian Lawrence Stone says:

> In the seventeenth century the early training of children was directly equated with the baiting of hawks or the breaking in of young horses or hunting dogs. These were all animals which were highly-valued and cherished in the society of that period, and it was only natural that exactly the same principle should apply in the education of children.[66]

Governesses, who were quite likely to end up as frustrated old maids, seem to be a breed well equipped to carry out this task of crushing the will and thereby snuffing out the evil in their charges. They had a reputation for malice and ill-temper. 'If an old maid should bite anybody, it would certainly be as mortal as the bite of a mad dog,' remarked Daniel Defoe.[67] As a result of the shortage of suitable males,

> Due to the low level of nuptiality among younger sons and to the rise in the cost of marriage portions, there developed in the eighteenth century a new and troublesome phenomenon: the spinster lady who never married, whose numbers rose from under five per cent of all upper-class girls in the sixteenth century to 20 or 25 per cent in the eighteenth century.[68]

The only possible occupation for these well-educated spinsters was to turn governesses. They suffered from economic hardship and social stigma. 'Not a relation, not a guest, not a mistress, not a servant, the governess lived in a kind of status limbo. By reason of her position, she was also treated as almost sexless. Not a lower class servant and so open to seduction, not a daughter of the house and so open to marriage offers, she was nothing.'[69] Being virtually shut out of society she took out her frustration on the children. Shocking but true. Fiends could go on masquerading as governesses for years on end in upper-class Victorian households without the parents becoming any the wiser. Besides, there was the deep-seated belief that to spare the rod was to spoil the child.

But what raises serious doubts in this case is the character and behaviour of Lord and Lady Scarsdale as now revealed. In their care and anxiety, they display a concern that is astonishingly modern. No governess, however Jekyll-and-Hyde-like, could have hoped to escape the fussy perseverance that they showed in the upbringing of their eldest son. Take, for instance, the time before sending him to his first prep school. One whole year before sending him the Scarsdales track down a parent, a Mrs Philips, who has a son there and badger her with questions about the school.[70] The contents of the letters show that every minute detail concerning the child's well-being is discussed threadbare. While assuring the anxious parents that there was also a nurse on the premises, Mrs Philips tells them, 'one thing we think is that the boys are obliged to bathe in cold water all the year round.'[71] Blanche Scarsdale is also in regular correspondence with the Headmaster's wife, Mary Powles, who clucks sympathetically, 'It is no wonder that every mother should feel anxious on first parting with her child.'[72] Surely such a mother was incapable of tolerating any form of violence to her children?

What also makes the story doubly surprising is the disposition of the ten year-old boy as seen through the letters written by him to his parents. They scarcely reflect a boy who has passed through a miserable childhood. If he had indeed, 'cried so much and so justly,' his subsequent behaviour certainly does not show it. If he had been subjected to the soul-shattering that has been claimed, he could hardly have become the breezy, confident child we see before us. The ten year-old Curzon who is supposed to have emerged from a three-year reign of terror in the nursery appears quite unscathed. The letters written from Wixenford show that his spirit is buoyant and his confidence boundless. Soon after reaching school he drops and breaks his watch, a very precious commodity in those times. But unlike a child fearing harsh punishment, he casually informs his mother about the incident and asks her in a businesslike manner about where he has to send it for repairs: 'So in your next letter, will you tell me if I am to send it to Leroy [?] and all about the directions and registering.'[73]

From school he is peremptory in his requests for hampers of food, seedlings, stamps, extra pocket-money. He asks for presents not merely for himself but also for his friends and lays down exacting instructions about how they should be selected. His mother always obliges, even going out to get a penknife to be presented to a friend. He told her, 'It should be of ivory and flat...not round as some knives are but as flat as the small pencil knife or the one we cut pencil with,' and it had to be purchased only from Ratcliffe's.[74] Curzon was eleven when he wrote this letter, and bounces with the confidence of a child who knows that his request will be met. Parcels from home arrive and he exultantly records receiving, 'netting things and beautiful warm things, splendid balls'. Besides, there is—'jam, pots, grapes, pineapple and cakes, apple and jelly alright . . . the cake looks a beauty.'[75] Such parental attention may be more expected if Curzon was the only child or perhaps the only son but he was one of nine. Blanche's fourth and youngest son, Assheton, was then only two and daughter Elinor was born the very year Curzon left home for Wixenford. Blanche was to give birth to yet another daughter, Geraldine while he was still there. Yet this did not diminish her time, attention, or devotion given to her eldest son. Curzon, for his part, adored his young mother, confessing to her his unmanly homesickness. 'I felt leaving you much more than I did when I first came to school, for I cried in bed last night, and was ready to cry every minute.'[76] He told his parents of the shocking greed of Mary Powles, the Headmaster's wife who seemed to have been in the habit of gobbling up the children's food. 'I think Mrs Powles must have taken nearly all my goodies away for there are only a few left.'[77] He had thus no fears or inhibitions about calling his Headmaster's wife a thief in a letter which may well be read by the master, not an unusual practice in boarding schools. Yet he expects us to believe that in his own nursery at home, surrounded as he was by other Scarsdale children, he remained mute!

We have only Curzon's own version of the governess. None of the other Scarsdale children seem to have left any account of her behaviour.[78] There are no available records of her antecedents. Mr Kenneth Rose, who has written a fascinating study of Curzon's early life, says that he reached a blind alley when he tried to trace Paraman's roots.[79] Indeed, Miss Paraman seems to have had no one she could call her own, for she left whatever little money she had to Curzon's elder sister Sophy.[80] Significantly, Blanche Scarsdale had a kindly opinion of her, as Curzon's younger brother Alfred records in his diary soon after their mother's death in 1875: 'Tuesday, 20th April. Very fine day. Papa gives Miss Paraman a gold necklace of dear mama's which she said she wished Miss Paraman to have.'[81]

Interestingly enough Curzon's description of Paraman appears in his *Notes on Early Life* and seems to have been especially written for a biographer in the years after his tenure in India.[82] Perhaps Curzon was looking for a scapegoat for the humiliating end to his viceroyalty and the lonely wilderness years that followed. In India he had set himself a lofty pace, fashioning a pattern of behaviour for himself based on his past successes, but the actuality could hardly be reconciled with his grand dreams. The result had to be accounted for and explained away, if not for others, for his own peace of mind. Tormented by failure, was he then driven to myth-making; making himself out to be more sinned against than sinning?

We know that in later life Curzon did disparage his father's role in his life and career. He made himself out to be a self-made man who enjoyed little paternal support or sympathy, telling his friend Lord Riddell, 'I may say I am a self-made man . . . I had to fight my way in life. . .'[83] Lord Scarsdale might not have had the means and the influence to further his son's career, but as we have seen he followed it with great interest. A strong bond existed between father and son. In fact, so strong were the ties that Curzon was prepared to risk his wife Mary's displeasure by coming to his father's defence. Once, when on a trip to England from India, Mary had complained of neglect by the Scarsdales. Curzon had promptly written back to say, 'I am sorry you think my people have not been quite what they ought to have been. I am sure that my father has nothing but the deepest affection for you and is incapable of intentional neglect.'[84] Such examples show elements of contradiction between Curzon's claims and the actuality, and so point to an attempt by Curzon—however subconscious it may have been—at distorting the truth.

Chapter 4

Golden Years
at Wixenford and Eton

In May 1869, Curzon went to his first school at Wixenford in Hampshire. He was ten, not an unusual age for boys of his background to be sent to boarding school; Lord Salisbury, thrice Prime Minister, and Winston Churchill both had left home at an even earlier age. Before going up to Wixenford, Curzon made a ceremonial leave-taking. He wrote: 'Dearest papa, Mama & grandmama. I invite you to come and see my little leave-taking rememberance (*sic*) to those I love, & to others I feel interested in & I hope you will like them. Will ten. 30 suit you? Please say. Always your loving child George.' A toast was raised in his honour by the staff on this occasion, for along with this letter is recorded on a sheet of paper in Curzon's writing: 'One shilling for John Strelton—William Thompson and John Robinson, to drink my health, wish me success, & I don't think they can get intoxicated with it.'[1] The adoring parents obviously thought the event of their son's going to his first school worthy of celebration.

Blanche personally escorted him to school. Curzon recalls, 'I remember to this hour the horrible moment when I saw the fly and white horse drive away carrying my mother, who was dearer to me than anyone in the world.'[2]

The school was expensive, costing as much £150 a year per

child, but this fact did not seem to have deterred the parsimonious Lord Scarsdale.[3] There, the Headmaster Revd R. Cowley Powles, was an 'amiable' and 'perfèct' gentleman, but 'weak' and 'querulous', and it was a second master, Mr Dunbar, who was the key figure. 'Powles did not associate us very greatly himself owing to the masterly influence and predominance of Dunbar,' Curzon said, 'but we were always delighted when he walked or talked with us especially when he talked of Kingsley or his Oxford days.'[4]

Training in imperial attitudes came from a maybe unexpected quarter at Wixenford. The Headmaster had moved his school from Blackheath to Hampshire to be near his old friend, the Revd Charles Kingsley. Curzon wrote to his mother: 'We have just come back from Church where Mr Kingsley preached; he preaches so simply, so that all may understand.'[5]

In 1886 Charles Kingsley declared he no longer subscribed to the idea of the equality of mankind:

Nearly a quarter of a century spent in educating my parishioners and experience with my own and others' children . . . have taught me that there are congenital differences and hereditary tendencies which defy all education from circumstances, whether for good or evil . . . I have seen also, that the difference of race are so great, that certain races, e.g. the Irish Celts, seem quite unfit for self-government and almost for the self-administration of justice involved in trial by jury.[6]

Charles Kingsley laid great stress on the importance of hereditary traits in forming character. No amount of education or training could really change them. Kingsley used Darwin's argument of the survival of the fittest in order to justify British dominion over other races:

Physical science is proving more and more the immense importance of Race; the importance of hereditary powers, hereditary organs, hereditary habits, in all organised beings from the lowest plant to the highest animal. She is proving more and more omnipresent action of the differences between the races.[7]

Charles Kingsley, along with Ruskin, Carlyle and Tennyson

had defended the action of the Jamaican Governor, Edward Eyre, for ordering the death penalty for the black politician G. W. Gordon and hundreds of other rebels in the Jamaican crisis of 1865. Upon Eyre's return home in 1866 Kingsley praised him for representing the 'English spirit of indomitable perseverence, courage and adventure.' Martial law, passed with Eyre's sanction, had resulted in the killing or hanging of 439 persons. John Bright, J. S. Mill, Spencer, and interestingly, Darwin, had opposed Eyre's actions.[8]

Curzon wrote regularly to his parents from Wixenford. School began at seven and, with breaks, went on until five. In the evenings the boys often played music. Bedtime was at eight. Curzon said, 'I am as happy here as I could expect to be, but I miss home very much.'[9] The Scarsdales often visited him. He began one letter to his mother saying, 'I should not have much to tell you today as you have come here the other day.'[10] She brought him a hamper of goodies which he generously distributed to most of the boys, though they did not quite go completely around the school.[11]

The Headmaster, Revd Powles, wrote a glowing account to Lord Scarsdale: 'Nothing can be more satisfactory to me than your son's conduct throughout the whole of his school-term. He has been uniformly industrious, obedient and well-mannered.'[12] Comparing Curzon with his younger brother, Alfred, who had joined him at Wixenford, Revd Powles said, 'Alfred has not George's intelligence.'[13] But it was the second master, Dunbar, who made a great favourite of Curzon. In Curzon's words Dunbar was, 'a short, stout gentleman with a moustache, whiskers and a little beard;—as a master he was for the most part detested by the boys to whom he was savage and cruel.'[14] Nevertheless, Dunbar seems to have had a soft spot for Curzon.

Flogging by masters, far in excess of complaint, was a normal part of a schoolboy's experience in sixteenth and seventeenth-century England, a bundle of birch rods was considered an essential part of a school master's equipment. Flogging covered various lapses, whether academic stupidity, disobedience or lying. The widely accepted procedure resorted to was to lay the child on a bench and to flog his naked buttocks with a bundle of birches until blood flowed.[15] Within days of first reaching school, Curzon was reporting to his parents, 'Gaskell has been caned twice and two other boys once: they have a whack on each hand for no great offence but more for disobedience.'[16] Powles also believed that to spare the rod was to spoil the child. Curzon recalled, 'He used to cane sometimes but very rarely on the hands leaving that as a rule to Dunbar and I remember once he swished me but for what offence I have not the least recollection.'[17]

Homosexual or sadistic motivation behind such flogging cannot be ruled out. A pamphlet published in 1669, cited by social historian Lawrence Stone, tells of some victimized boys who complain, 'our sufferings are of that nature as to make our schools to be not merely houses of correction, but of prostitution, in this vile way of castigation in use, wherein our secret parts ... must be the anvil exposed to the immodest and filthy blows of the smiter.'[18]

Beatings were justified as being the only method of inducing learning. But the accompanying savagery indicates that for the masters it may have also provided a release for their own frustrated libidoes. Many of the masters remained unmarried all their lives. Dunbar died a bachelor. The relationships between Dunbar and Curzon was such, however, that Curzon was able to report to his mother, 'Mr Dunbar was pleased with my bottle of scent.'[19]

Flogging was not always resented by the boys and Curzon does not seem to have minded it unduly. He says:

> But comic to relate I still remember the delicious feeling of warmth that ensued about 5 to 10 minutes later when the circulation was thoroughly restored and the surface pain had subsided. With the birch I think he [Dunbar] never gave beyond ten or twelve strokes— and that for some particularly grave offence in his bedroom at night.[20]

As he reached puberty, Curzon began to attract the attention of people with established homosexual tendencies. Their overtures were more mental than physical. At Wixenford there was Dunbar. At Eton there would be Oscar Browning.

Curzon recalled how Dunbar ' . . . could be extremely nice to us when he was in a gracious mood.'[21] Dunbar brought his good-looking pupil many presents. 'Isn't this a pretty transferable?' asked the ten-year-old boy of his mother, fixing it on the letter. He mentions, 'Dunbar got it for me in London and some others.'[22]

Dunbar's whipping did not arouse lasting resentment. Curzon was proud of the attention he got from the powerful master and strove hard for it by rising to be head of school. In his last term he carried away five prizes. Curzon was elevated to the rank of the school treasurer:

> I also under Dunbar's supervision kept the school account...I still have the tin cashbox in which I kept the money and the account books...I believe I was never out by even a penny at the end of the term, and undoubtedly the plan was a wise one as inculcating both business acumen and economy.'[23]

This habit of his childhood never left him. First manifested at Kedleston at the age of seven, he records in his diary the purchase of a kaleidoscope for one shilling, a string barrel for four shillings, and a leather purse for one shilling.[24] It was that same fastidious trait that made him keep a catalogue of his expenses on his first Grand Tour, in 1886. '5 days Hongkong £4-5-0 . . . 31,500 miles at a cost of £1.15.0 a day.'[25]

Like many upper class boys, Curzon went to Eton. Founded by Henry VI in 1440, on the banks of the Thames facing Windsor Castle, Eton offered, apart from its curriculum of studies and games, much extravagant freedom and whispered delights.

Eton was the surest passport to a successful career, whether in politics, the Church, diplomacy, or the civil service. Those who went to Eton learnt to view life strictly as comprising a small select aristocracy destined, as it were, to rule a large mass of people born to be ruled. The unspoken rules governing Eton were those of the most exclusive and most influential clubs. It was the acknowledged nursery of Britain's Prime Ministers. Gladstone was an old Etonian, as were Salisbury, Rosebery[26] and Balfour.[27]

Until the arrival of Headmaster J. J. Hornby in 1867, boys at Eton continued to be educated primarily in the Classics. Hornby, an Oxford Blue and a well-known athlete, broke the stranglehold, adding Modern Languages, Geography and History, in an attempt to bring Etonians to a more realistic awareness of the world around them. But it was a slow, uphill task. As one social historian has said:

> The public school was physically as well as spiritually withdrawn from the world behind its high gates and walls, looking inward to its own stretch of lawn and its haphazard and yet harmonious architecture, Gothic and Georgian and neo-Gothic... Here in sombre religious shade, the future rulers of England heard their headmaster preaching about honour and service and sin.[28]

The growing threat posed by the other European powers to British predominance did not disturb the placid calm of the public schools which cherished the myth that the English were a people with a special mission.

In going to school, 'The young aristocrat exchanged the lawns of his own house for the equally fine grass of his school quadrangle.'[29] Mathew Arnold bemoaned, 'The society of a public school is a world in itself, self-centred, self-satisfied. It takes but slight account of the principles and practices which obtain in the world of men.'[30]

A victim of these influences, brilliant and hard-working though he was, Curzon was not able to look outside or beyond the ivory tower of Eton. It reinforced his belief in the superiority of good breeding, tradition and education and perhaps because of this, he tended to view with near contempt 'the principles and practices which obtain in the world of men'. In the two crises of his political career—in his humiliating fight with Kitchener and in his frustrated aspiration to Britain's premiership—he was outwitted by those whom he had always regarded as men of inferior pedigree and education.

Nevertheless, Curzon maintained that the happiest years of

his life were spent at Eton.[31] He went to Eton soon after his thirteenth birthday. Like all great fifteenth-century buildings, the school was built around various quads; the outer one housing the chapel, and the Upper and Lower schools, with the bronze statue of their royal founder dominating the quad itself. The inner quad, or the cloisters, contained the residences of the officials, the library, the dining hall and the offices. Classes were known as Divisions and contained as many as 100 boys; every boy was assigned to a Tutor, generally his Housemaster, with whom he lived.

Curzon's Housemaster was Wolley Dod, a tall, thin, querulous man with narrow reddish whiskers.[32] Unfortunately for Curzon, Dod was not prepared to treat him as a great personage. Hitherto, whether it was his parents, uncles, aunts, or the masters at Wixenford, all had combined to make George Nathaniel Curzon feel that he was a special person. Now, to his chagrin, he found that at Eton the teachers were not prepared to automatically pay him the homage he had begun to believe was his due. It was an unusual experience for him and not a pleasant one at that. Curzon was deeply incensed and the conflict between what he was and what he felt he had to be seems to have endangered his identity, releasing potent forces.

Curzon geared himself to devise his own special way of 'scoring off' against the masters. He provoked them to have him thrown out of class. Once in his room, however, he spent half the night in study in order to win through private exertion the prizes other more favourite pupils were expected to win and to thereby experience a kind of 'sweet revenge' in this one-upmanship. He boasted,

> The vein of devilry on my part and of blunt obfuscation on that of the masters continued throughout my Eton career. They never could realise that I was bent on being first in what I undertook, but that I meant to do it in my way but not theirs.[33]

It is generally believed that neglect and cruelty during childhood find an outlet in rebelliousness later. Curzon's defiance at Eton, however, can be interpreted as the result of his fury at not being made to feel special. Significantly, at Wixenford, Curzon had felt no need to revolt as the masters had fussed over him, offering much attention. At Eton, the two French masters in particular, according to Curzon, regarded him as an impossible pupil, which whetted him to winning the coveted Prince Consort's French prize. He did win the prize, as he put it, 'by a larger percentage of marks at an earlier age than had ever been done before.'[34]

Curzon carried his indiscipline and independence to wild excesses because of the sense of power he felt at being able to do exactly what he liked. He indulged in reckless action, which was often dangerous and foolhardy, for no reason other than the sense of control it gave him. He confessed:

> I made it a point of honour to attend Ascot Races every year not because I cared in the least for racing but because it was forbidden and therefore dangerous...Another of my somewhat daring eccentricities was that I had a zinc lining made for the bottom drawer of my oak bureau, in which I used to keep a stock of claret and champagne. It was not that I cared for drinking but that I enjoyed the supreme cheek, as an Eton boy, of giving wine parties in my room.[35]

It was scarcely surprising that this bold defiance made Curzon one of the most popular boys at school. His contemporaries gazed with awe at his reckless daring; even the older boys were stirred to take notice of the handsome, heedless youngster. Wolley Dod complained to Lord Scarsdale:

> Being young for his place...and a popular and well-mannered boy, he is in some danger of being spoilt by associating too much with boys older than himself. I believe him to be very well principled, but the notice of older boys is too apt to make young ones conceited and forward.[36]

Lord Scarsdale seems to have turned a deaf ear to the complaint. Dod's tone rose to a fever pitch in a later tirade. He said of Curzon, 'I cannot let him have his own way in the way he fights for it,' adding: 'I am as long suffering as most of my colleagues.' It was necessary, the livid Dod emphasized, 'to be very peremptory' with Curzon for he had 'questioned my order and my authority in what I call an intolerant and unbecoming manner.'[37]

It is interesting to contrast Dod's complaint with the report of the earlier Master, Revd Powles, who had praised Curzon for being 'uniformly industrious, obedient and well-mannered'. But then, Powles, unlike Dod, never clashed with his pupil. 'Nothing can be more satisfactory to me than your son's conduct throughout the whole of his school-term,' he had said.

But Lord Scarsdale remained mute to Dod's complaint, unable to bring himself to upbraid the son who was the apple of his eye. This becomes more evident when the exasperated Dod forwards to Lord Scarsdale a letter from another long-suffering master, H. G. Madan, who catalogues his complaints against Curzon:

> My dear Wolley Dod, I am afraid I ought to write to complain of Curzon for constant talking and interruption in school. He has an irrepressible habit of asking foolish and what I call 'chaffy' questions. I threatened to send him out of the room, and he said, 'he did not care'. He seems utterly wanting in respectfulness, a pert, restless and sharp little child.[38]

We do not have Lord Scarsdale's reply to Dod but it appears that he preferred to defend rather than condemn his son's behaviour. This is revealed by Dod's reply to Lord Scarsdale in which the master sheepishly acknowledges that the failure may have been on the master's side: 'I admit that several of the extra masters not being schoolmen themselves [including Mr Madan] do not rightly understand the treatment of boys.'[39]

In putting Dod on the defensive, Lord Scarsdale once again demonstrated his extraordinary partiality for his eldest son. It was the master who was to blame for he did not rightly 'understand the treatment of boys'. There seems to be little in the boy's behaviour that the father found worthy of reprimand. In the letters to his son, nowhere does the stern Lord Scarsdale admonish his son or take him to task for his increasingly impertinent behaviour and bouts of defiance in class. In fact, in his last year in school, Curzon cajoles his father so as to have him transferred from the clutches of Dod to that of another master, E. D. Stone. Surprisingly, Lord Scarsdale complies and Curzon boasts with supreme self-confidence to his fairy godfather, Oscar Browning how 'My father saw Dod in the holidays and told him he thought he did hardly enough for me; so he himself effected the exchange.'[40]

Under Stone things were no better. Curzon had got embroiled in a headlong clash: 'Almost immediately afterwards I had a row with Stone who strutted about his pupil room telling me I had not a single gentlemanly feeling in my head.'[41] Despairingly, Stone said of his wayward pupil, 'He has marvellous quickness and good memory and if only he could concentrate his faculties more he would do thoroughly well.' But Stone found Curzon was 'far too superficial to do anything which requires deep thought, and energy of purpose,' and strongly believed that 'disappointment may be the best school for him.'[42]

In saying this Stone was being prophetic. His effortless ascent up the ladder of success at an early age was to prove Curzon's undoing. But unfortunately, the overtly fond father was not able to chide or control his son and thereby protect him from the destructive power of his yet untrained will. Tragically, while encouraging his son to express his individuality, Lord Scarsdale failed to provide the safeguards that would shield him from overreaching himself. Unfettered and unrestrained, the autonomy of the will was to turn the boy against himself.

It is thus one of the supreme ironies of Curzon's life that the care and concern of his parents should have paved the way for his undoing. His doting parents made him feel he was so superior that he had to carry the burden of his admirers' ambitions. In an era which generally believed in crushing the libido, his parents had showed no hostility to their child. Tragically, their permissiveness put another burden on his shoulders, their love and attention not merely becoming models for fulfilment but also paving the way for disaster. Out of a brood of eleven children, he was singled out for special attention and he had to bear the burden of this responsibility. Curzon thus grew up only being at peace with himself when living up to the goals expected of him; goals which were not always consistent with the timetable of his needs and the chemistry of his body.

Of course, Lord Scarsdale did occasionally preach to his son on the sins of vanity. While congratulating him on winning the Prince Consort's French Prize, his father warned, 'Do not celebrate your success by wearing your hair long and wrapped round your ears! You know what I mean and I do detest long hair.'[43] From time to time he also imposed strict economy. Clothes had to be worn until they were threadbare. Curzon had to beg for a new pair of trousers for 'the holidays and next half, as mine are rather shabby now and I have not had a new pair this half.'[44] He had also to plead for a new Eton jacket. 'May I have a new Eton jacket?' he had written to his mother. 'I have only got two and one of them is so shabby that I cannot wear it. So I have only got one fit to wear; I use that on week days and have none for Sundays. The two jackets which I have now I have had for more than two years so I have worn them well. If I may have a new one please will Papa send me an order or tell me I may as I really want one.'[45]

But apart from these minor irritants, Curzon was allowed to live like an aristocrat at Eton. His room was reported to be one of the more opulent ones and Gladstone on visiting Curzon there in 1878 professed 'to be aghast at the luxury of pictures and China, armchairs and flowers, compared with the plain living of his own day.'[46] Curzon's mother had helped him decorate it. When she sent him two China pug dogs for his mantlepiece, Curzon had delightedly written back, 'I have quite a menagerie now and if I get two black elephants it will be quite a show of Derby China.'[47] Purple damask curtains draped the windows of his study. He had consulted his mother about the colour and expense, asking, 'Would Papa mind paying half of it?'[48] As expected, Lord Scarsdale had obliged.[49] Earlier Curzon had asked for a new writing table: '. . . they are 15/-, £1, £1.5.0 here. May I get one or had I better pay for it with my own money?'[50] Paternal generosity had once again prevailed. 'Please thank

Papa very much,' he wrote three days later, 'for letting me have a writing table. I have ordered one and they will send you the bill.'[51] Curzon was also given his new evening tailcoat, though somewhat grudgingly, and then only after he had won the Prince Consort's French Prize. The frugal habits of a country vicarage died hard and Lord Scarsdale wrote, '... before ordering an evening tailcoat do tell us if other boys of a similar age and height *commonly* have them and if you really consider yourself uncomfortable without one. I quite thought your black first tails would suffice for the present at least.'[52] A black tailcoat, white bow-tie and a top hat were compulsory attire for the boys at Eton and it was not until the Second World War that top hats were abolished; younger boys wore jackets with a black tie. But while Curzon seems to have had to plead for clothes, he had an allowance which was generous enough to pay for the stock of claret and champagne in his room, a facility not enjoyed by most of the other boys.

Much can be made of Curzon's repeatedly turning to his mother, soliciting her help whenever he wanted a favour from his father: Here, surely, are shades of an Oedipal complex. In the conventional Victorian household the father appeared by far the most dominant member. In his presence everyone else, including the mother, seemed to diminish in size. The easy intimacy which a boy shared with the mother was pushed into the background in the presence of the father. Boys grew up feeling that they could capture the attention and admiration of their mother only in the absence of their father. Unconsciously, the mother helped to perpetuate this myth by indulging them in private and acting as a buffer against the father, whom she presented as an embodiment of authority.

As is seen in Curzon's growing years, his mother did foster a feeling of special intimacy with him and he felt he could approach her with demands he could not so easily put to his father. But, it is also seen that though he found it easier to communicate with his mother, Curzon did not go out of his way to exclude his father. During the storms of adolescence, Curzon seems to show little problem in adjusting his growing identity with the paternal image. He is seen to be able to face his father squarely, displaying neither postures of open defiance or too submissive an obedience. He is able to take Lord Scarsdale's pecuniary harshness and aloofness in his stride, conscious that it is balanced by tenderness and regard. In fact, Curzon's attitude shows he is aware that in his father's eye he can do no wrong.

Blanche was not able to visit her son at Eton but this did not diminish the deep affection between them. Her repeated pregnancies must have taken their toll on her health. Soon after going up to Eton Curzon wrote to her, 'I was very glad to see Papa on Thursday but I was so sorry that you couldn't come.'[53] Blanche's letters to her son at Eton are few and far between. But they express intense pride in her eldest son's achievements there. Upon Curzon's winning the Prince Consort's French prize, Blanche writes to him what can only be described as an exuberantly gushing letter:

My dearest George, Papa came into my room at 9.15 or so this morning holding W. Dod's open letter in his hand and saying 'such good news from Eton, G. has taken the French Prize!'

W. Dod has written so very kindly and nicely—I am sure he is as pleased as anyone—and it is a great mark of distinction for you at your early age—and I only hope your shiverings and ailings don't your over tacking (*sic*) your brain—and consequent depression and bodily langour—It may be partly cold too—

Papa has written off letter to W. Dod—Oscar B, G. Mama C. This is all I think, and I am writing to H. Hall & Bray, it will be in the papers of course. I *hope* will be rightly spelt—I wonder if the Queen sends for the boys, I hope so. There are 2 prizes, or rather *first* and *second*. I suppose—and I fancy only 20 or so compete, and were you not 6 or 8th among that, 20 or so in 1873? I can't think why Smith's name is put above Broderick's in prize men list 1873! In School lists you sent me Michalmas on the 4th leaf Page 8 of the book are the lists

French	German	Italian
Smith	Sargeant	Crowder
Broderick	Munday	Halloway

& page 13 is named at bottom as taking Prince Consort: first French Prize 1873 so he ought to come first.[54]

On the left-hand top corner of the letter Curzon has written in pencil, 'very very precious'. The letter shows that not merely was Blanche proud of her son but that she followed his school career with intense interest, even remembering in detail the progress of his classmates. Young Curzon never forgot his mother's pride in him and felt he had to strive throughout his life to maintain her high regard. He was always to be the boy the Queen would send for, this had been Blanche's wistful hope for her first-born son. Exactly four months later she was to die of typhoid. Curzon never forgot her hope and in all his actions thereafter he kept looking over his shoulder for her approval, trying to live up to the goals she had set for him.

So great was the pressure to excel that he took shortcuts to maintain his position. In the summer of 1875 Curzon's tutor, Wolley Dod, wrote to Lord Scarsdale complaining that his son

... was guilty of taking a leaf out of a book into 'Collections' and using it in the examination. I am sorry to say that the standard of morality about copying work 'cribbing' as it is commonly called is not as high as we would wish it but it is treated as being a serious matter.[55]

Circumstantial evidence shows that three years later Curzon was again found cheating. Oscar Browning came to hear about it in Cambridge and expressed his sympathies. Curzon wrote back:

It is very kind of you to say what you do about 'that disgraceful scandal'. It has been a very painful business to me as you may imagine and I am afraid threw me back a place or two in the select. But I am glad it was most satisfactorily settled at Eton (except possibly in the minds of those who haven't heard the whole story) and feel sure that I can satisfy doubts in anybody if they exist. You shall hear the whole history and will I think approve of the action I took. I had a long talk about it with Cornish at Eton and it pained me very much that he did not seem quite to believe me but the strong and unanimous assurance of those to whom the matter was referred to for decision that not *the slightest* imputation could be cast upon my honour and that my explanation was in every sense a refutation, added to the kind belief of true friends like you and my own confident assurance—tender the somewhat callous to doubts and insinuations.[56]

Years later when Browning visited Curzon in India in 1902 he told Walter Lawrence how Curzon 'was accused unjustly of being a copyist.'[57] Browning seems to have been recalling this episode. In a letter to Browning, Lord Scarsdale refers to Curzon's disappointing performance in the Select. But the father does not seem to be aware of 'the disgraceful scandal'. He begs the master not to let Curzon know of his disappointment:

I confess to a slight feeling of disappointment at him not standing higher in the list—but you must not say this to him, please as he has worked very hard and seems quite satisfied. The Select were all very close upon each other in ranks, he tells me and I have no doubt his examination was very creditable.[58]

Curzon rose to become President of the Literary Society and auditor of Pop, Eton's most exclusive debating society. In the summer of 1878 Curzon invited the Liberal leader Gladstone, then out of office, to come and address the Literary Society.[59] Though not a Liberal himself, Curzon admired Gladstone as a man. Besides, having an ex-Prime Minister as a chief guest was a feather in his cap and Curzon was not one to shun such dividends.

A die-hard Tory from the outset, Curzon had evinced strong interest in the General Elections of 1874 that brought the Tories with Disraeli as their leader to power. Though he had then barely finished a year in school Curzon and his friends disobeyed the Headmaster and had gone to Windsor to see the polling. 'I with a lot of other fellows managed to evade them [the masters] and we slipped into Windsor; the master pursued us and we ran in and out of the Castle,' he told his father.[60]

In 1869, the year ten year-old Curzon left home for his first school, Disraeli sent Lord Mayo to India as Viceroy.[61] Mayo had begun energetically, taking the white man's burden seriously, maintaining the efficient administration which was thought to be the primary justification of British rule in India. But before he could finish his term, he was murdered by a Muslim tribesman who was believed to be in conspiracy with the larger unrest that was spreading across northern India. A frenzy of rage enveloped the British community as a repetition of the horror of 1857 was envisaged. Sir James Fitzjames Stephen, a member of Mayo's Council in India, seems to have been particularly shaken. He told his wife hysterically that if the high Court acquitted Mayo's murderer, he would hang him himself.[62] Upon his return to England, Fitzjames Stephen lectured at Eton and Curzon's imagination was fired by the India he described. Years later, Curzon recalled how 'the fascination and, if I may say so, the sacredness of India' gripped him from that day.[63]

Curzon finished school in a blaze of glory. He was fêted at his final annual day celebrations and records the heady boost to his ego in proud but vulnerable tones, apologetically saying, 'this is decidedly egotistical'. So it was, but most endearingly so:

June 4 1878. The proudest day, I *expect* of my life. Speeches in the morning. Remarks overheard...
'Where is Curzon? I heard him last year and liked him so much.'
'Isn't he splendid?'
'What wouldn't I give to be his sister?'[64]

Lord Scarsdale took pride in the achievements of his eldest son. *The Daily Telegraph* of 5 June 1877 records his presence at the annual day ceremonies at Eton. The paper proceeded to single him out for special praise: 'Mr Curzon, who not only speaks with dramatic power, but acted the various characters he assumed to a T, and displayed an intimate acquaintance with French & Italian.'[65] The proud father sent him a cheque for £5 after returning from the celebrations at Eton. In his letter he wrote, 'thoughts of your darling Mother, were uppermost in my heart, of that she could have been at my side.'[66]

On 4 April 1875 Blanche, barely 38 years-old, had died of typhoid. The Scarsdales had assembled in London for the Easter holidays when, all of a sudden, Blanche took to bed with a sore throat and a headache. Soon it was clear that Blanche was suffering from typhoid. Her distraught husband summoned the best available medical help, no less a doctor than Sir William Jenner who had attended on the Prince Consort in 1861.

But all efforts proved futile. A week later, Curzon sadly wrote, '... dear mama died at half past three quite peaceably, with no pain. We went in and saw her today, her face was like it used to be, with a happy smile on ... Aunt Mary put white flowers all over the bed. We shall not see the dear face again.'[67] Curzon picked up one of these flowers and preserved it for the rest of his life. It remained amongst his most precious possessions.[68]

Tragedy continued to dog Curzon's footsteps. A severe injury suffered from a horseback fall during holidays at Kedleston when he was only fifteen, and forgotten after a few days rest in bed, now cropped up again. It was to torture him for the rest of his life, turning him into a highly-strung human being. It has been said that human behaviour emanates from a combination and permutation of the psyche, the body and environmental influences; thus this physical injury of Curzon's is important in understanding the full man.

On the eve of his entry to Oxford, Curzon was holidaying in France where, he wrote to a friend, 'I have felt shooting pains in my side—in the region of hip and noticed the unusual prominence of that member.'[69] He saw a physician in London who diagnosed curvature of the spine and told him to abandon all hope of going to Oxford. Curzon, however, was determined to find a way out. He went to the Harley Street specialist Paget who held out hope. His report said:

> There is strong reason to expect that the increase of curvature which no doubt, so far as any had occurred, has been directly due to overwork—will be corrected through favourable conditions which can now be secured.[70]

Considerably reassured, Curzon reported to his master Oscar Browning,

> Paget I am thankful to say modified the opinion or rather the Decision first expressed. He has given me permission to go to Oxford ... but I am to go as an invalid, ie. I am to wear perpetually [a] steel appliance which is now being made for me ... I am to lie back in a specially made chair a large portion of every day.[71]

The incessant throbbing pain only drove Curzon to work

harder. 'Owing to my infirmities—my bad back and my bad leg—during recent years I have been more and more driven to work, which has enabled me to fight the pain which I almost constantly suffer.'[72] But overwork intensified the pain and so the vicious circle continued. Henceforth, his system was permanently affected. The pain in his back was perpetually there in the background. The great expectations that his family and friends—and Curzon himself—had of his career forced him to make a superhuman attempt to overcome this setback. He worked harder than ever, trying to be the superior man people had began to believe him to be.

Every morning he had to be fitted up in a steel waistcoat which he was to wear for every one of his waking hours for the remaining forty-seven years of his life. This steel cage gave Curzon a ramrod erectness of stature which added to the general impression of stiff-necked arrogance. Curzon said:

> My reputation is due to some measure to the fact that for many years I have been braced up with a girdle to protect my weak back. This gives me a rigid appearance which furnishes point to the reputation of pomposity.[73]

What is remarkable is that in spite of being in constant pain Curzon took long and hazardous journeys to remote outposts of the Empire, making himself an expert on the East. Both as Viceroy and later as Foreign Secretary, he worked fourteen hours a day. The result was damaging to both body and mind. When the physical pain increased, he could rarely sleep without drugs. When his viceroyalty started souring in India, his detractors were not slow in accusing the Viceroy of being sick in both body and mind.[74]

Subsequent to his spinal injury Curzon began to display a masochistic tendency.[75] Curzon seems to have sought out, almost deliberately, the powerful, ruthless characters who would hurt and humiliate him. In various stages of his life, Curzon seems to have positioned himself so that he was invariably crushed, if not physically, then psychologically. Despite his spinal injury he insisted on submitting himself to a strenuous workload which brought him to the verge of a physical and nervous breakdown. He did not seem to mind the pain, as if it gave him a vicarious satisfaction, allowing him the dual pleasure of proving himself superior and at the same time permitting him to wallow in bouts of self-pity.

Psychologists maintain that a careful study of the phenomenon of sadism and masochism shows there is no real demarcation between the two and that traces of both were often found in the same individual.[76]

Thus inhibited by a crippling handicap from fully achieving mastery over himself, Curzon's subsequent rage had to find an outlet; it did so by alternatively showing harsh cruelty to others and to himself. Though Curzon was no sadist in the accepted definition of the term,[77] there are occasions when he certainly seems to have gone out of his way to hurt his loved one. The premarital five-year association with Mary is characterized by much tantalizing on Curzon's part. But Mary was not the only one to suffer. After her death in 1906, Curzon had an eight-year affair with the actress-writer Elinor Glyn. The manner of its termination reveals not merely a cruel lack of regard for the lady, but also a peculiar vindictiveness.[78] Curzon had first seen the lovely red-haired Elinor on the London stage after Mary's death. Her play *Three Weeks* was causing a delicious scandal because of its naughty scenes on a tiger skin. He was attracted, his appetite further whetted by the knowledge that another colleague was besotted by the lady. Never to let a challenge go by, Curzon set out to pursue her. It was scarcely a difficult task. For Elinor, the cold, aloof, intellectual statesman was the stuff the heroes of her novels were made of.

In fairness to Curzon, however, it has to be said that he had made it clear to Elinor that their affair could never end in marriage. But Elinor was living in Curzon's home, Montacute, when upon opening *The Times* to read about Curzon's appointment to Lloyd George's Cabinet, she was staggered to see the notice of his engagement to Mrs Grace Duggan. Elinor Glyn's grandson was to later write:

There had been no letter beforehand warning her of what was to come; nor was there any letter afterwards. If only there had been some word of warning, some word of explanation, it might have hurt less. Their passionate love affair had lasted for eight and a half years, and now it was severed by one public blow of the axe.[79]

With his second wife Grace, however, Curzon, well past his fiftieth year, displayed another side of the same streak. He seemed to draw morbid pleasure from the humiliations Grace afflicted. Once, early in 1922, when seized by a shooting pain in his right leg and ordered by the doctor to France for cure, he had begged for Grace to be at his bedside. But busy preparing for a charity ball at Lansdowne House, Grace had not found the time to go. Alone and racked by pain, he roused himself to write pathetically:

I lay awake last night and thought of each stage of your party— going in to dinner, coming out, the arrival of guests and the beginning of the dance, Gracie taking turn with the 'juicy bucks' and so on hour after hour.[80]

He almost seems to have enjoyed the thought of her enjoying herself at his expense.

Chapter 5

Eros Encountered

While Curzon was in his first year at Eton several older boys were beginning to take notice of this extraordinarily good-looking fellow who had the audacity to take on the Eton masters. The Lyttelton brothers, nephews of Prime Minister Gladstone, and senior scholars at Eton, along with St. John Brodrick, singled him out for attention. Little wonder Wolley Dod complained to Lord Scarsdale that his son was in some danger of being spoilt by associating too much with boys older than himself. This attention had acted to swell the ego. Brodrick at the pinnacle of his own career, recalled his first meeting with Curzon in a railway carriage that he was sharing with Alfred Lyttelton:

> Just as the train was moving a tall, breathless, pinkcheeked and well-groomed boy with black hair was shoved into our carriage. He was covered with shame at intruding on so great a personage as Alfred, but recovered a little when we congratulated him on having been pronounced winner of the Prince Consort's French Prize that day, which, as it happened I had won the year before. So reassured, he gaily entertained us until our arrival at Paddington. It was the first introduction of either of us to George Curzon, and from that day in 1874 began a friendship which lasted without shadow for nearly 30 years.[1]

It was inevitable that Curzon's youthful good looks would get him embroiled in controversy. In his very first year the joyous, rotund, epicurean Oscar Browning drew him into his fold. Curzon confessed to being drawn towards Browning because he 'gave encouragement and inspiration, hence my life-long attachment to him.'[2]

At Eton, two masters, Oscar Browning and William Cory had sought to introduce culture to their pupils. The puritanical and athletic Headmaster Dr James John Hornby, had watched their efforts with dislike and written them off as aesthetes. Cory lost his job for having written some passionate letters to one of his pupils, the son of a Bishop. Browning, however, had stayed on at Eton.

Twenty-two years older than Curzon, Oscar Browning was reputed to have homosexual leanings. In November 1854 at the age of 18, Browning confided to his journal:

> I remember well, when I was nine years younger than I am now that I could not help thinking how odd it was I was a man . . . I am here at Eton. I have been unluckily thrown among a set who do nothing but ridicule my peculiarities among them. I am, according to them, incapable of performing in any way the duties of public or private life.[3]

Handsome young boys enraptured him. Browning recalls:

> I saw a boy named Dunmore. I was struck by his eyes. I have been more so by his manner, everything about him. My wishes, my hopes and fears begin and terminate in him. I have found that he is a lord, but I loved him before.[4]

Life in Browning's house was a pleasant combination of high thinking and good living. Helping him run it was his mother, a spirited and a charming woman, who spoke both French and German and played the harp and the piano. Parents felt happy and relieved that their offspring should be reared in this atmosphere of genteel decorum. Artists and men of letters, actors and musicians, Ruskin and Solomon, George Eliot and Walter Pater, were frequent visitors, bringing with them an atmosphere of culture and refinement so dear to Browning. Theatricals used to be given by the boys in the dining room; professionals came from London to perform chamber music. 'Arundel prints hung on his walls. Morris curtains framed the windows . . . Amidst these surroundings, in the genial warmth of the Pre-Raphaelite movement, the boys lived and grew from boyhood to adolescence, from adolescence to early manhood'[5] and fell easy prey to the charms of Oscar Browning.

In schools where the sexes were separated and boys grew to manhood in an atmosphere of monkish seclusion it was not rare to see the sprouting of intensely passionate friendships. Intense romantic attachments flared up between senior scholars and innocent youngsters. Boys were known to send perfumed valentines and address each other in gushingly effeminate terms. At Eton such amorous activities did not excite the violent distaste that they might have elsewhere.[6] Victorian society did not uniformly repress early sexuality but tacitly allowed it cautious play in its great public schools.

The study of the Classics must have encouraged such ideas. The concept of boy-love was idealized by the ancient Greeks who believed that in its pure, romantic, intense form it was the most exalted of all human passions. Thus Epaminondas, an acknowledged lover of boys, could yet be an honoured citizen. Aeschylus, Sophocles, Alexander the Great and Caesar also had male favourites.[7] The military academies of ancient Greece were known to have encouraged such attachments, to counteract softening feminine influences and to promote heroism.

Oscar Browning belonged to the Aesthetic Movement, many of whose protagonists were his personal friends. He later said:

> Few people knew that the aesthetic movement which had so much influence in England from Ruskin to Oscar Wilde had as one of its characteristics a passionate desire to restore 'Greek love' at the position which its votaries thought it ought to occupy. They believed that bisexual love was a sensual and debasing thing and the love of male for male was in every way higher and more elevating to the character . . . I was interested, but did not agree with them, as I was at that time a school master. It was absolutely impossible that I should take their view of things.[8]

Not confining himself to boys of his own house, Browning also took Alfred Lyttelton[9] and his brother Edward,[10] distinguished athletes and nephews of Gladstone, under his wing. Under the garb of trying to alleviate their anxieties, Browning invited confidences from the boys, saying that he was one master to whom they could appeal without restraint.[11] Such proposals invited fierce resentment. The teaching staff looked upon these activities with suspicion, fearing a witch-hunt and a conspiracy to spy. For 'spooning', as it was called, was considered '... the one dark fear that stalks in the background'.[12] Headmaster Hornby had viewed Browning's activities with intense distaste and it was against this background that Curzon found himself the scapegoat in a battle between Hornby and Oscar Browning.[13]

Oscar Browning made the acquaintance of young Curzon in 1872 and confessed to having been impressed by him as one of the most brilliantly gifted boys he had come across. Browning took upon himself the role of a guardian angel to boys of promise, and had a low opinion about Curzon's master, Wolley Dod, his anxiety being increased in the following Michaelmas, when the Captain of Wolley Dod's house told him that he was deeply concerned at the companions with whom Curzon was associating.

In the middle of 1873, Curzon had reported to his mother:

> The other day Mr Browning took me out for a drive about 14 miles out into the country, to see the churchyard where Grey wrote his elegy, & his tomb there, & also the house in which Milton lived when he wrote his *Paradise Lost*. I saw the very room in which he wrote it—a little closet about as big as two of the nursery WCs.[14]

Nevertheless, the dangers surrounding Curzon were such that in the summer term Browning received a letter from a senior student who complained about Curzon being maligned as a result of having borrowed a rug belonging to Curzon. The senior student said:

> I feel it more particularly in this instance, as Curzon is a boy among many whose acquaintance for various reasons I should

much like to make, knowing of what superior quality he really is and how often he has been in a dangerous position here. But bearing in mind the state of feeling about these matters, I have, of course, relinquished the idea fearing the harm it might do to other boys.[15]

It is a testimony of the close relationship Browning had with the student that he should confide in him about such a delicate matter.

The borrower of the rug was the cricket hero Edward Lyttelton. Among the Curzon Additional Papers there is a tiny slip of paper in which Edward has written: 'My dear Curzon, I send you back your rug which I should not have asked for so coolly, but I had had enough standing up during the day and was much obliged for it. Yours very truly, Edward Lyttelton.'[16]

Though there is no explicit mention of Curzon having invited promiscuous male attention, something about Curzon aroused a passionate protective urge in the older boys and masters. They seemed obsessed with the dangers which they feared surrounded him and might sully him. Whether it was Wolley Dod or Browning or one of the older boys, each one seems to have felt the impact of his youthful good looks and went out of his way to help him preserve his innocence.

The Lyttelton brothers, Alfred and Edward, who were to become his steadfast and loyal life-long friends, were considerably senior to Curzon at Eton. Nevertheless, they seem to have been seized by a compelling urge to guard him. Proven scholars, the Lytteltons were also famed for their athletic prowess, Alfred being the Captain of the School. Just a few hours before leaving school, Edward wrote to Curzon asking for his photograph. He was then 19 and an established hero while Curzon was a 15 year-old youngster. Edward, conscious of the awkwardness of his request, goes to pains to provide an explanation in the letter:

Perhaps it may have seemed odd to you that I who have had so very slight a claim to be considered as one of your acquaintances should have asked so coolly for your likeness, and I should like to give you an explanation of this and other things.

I have heard a good deal about you at one time and another from

Oscar Browning who I have already known for a long time now, and always like and in many ways respected. What he told me and what I have been told by other people could scarcely fail to awaken in me a considerable interest in a boy of your position, surrounded by so many dangers, and needing at times a helping hand.

I have always been very observant of the various phases of life presented at Eton, and the more I observed, the more it struck me that in a case like this there were difficulties in my path which should warn me to be careful. I well know the sort of view which the world takes to any big fellow hemmed in by the social chains of being what is called a swell, taking any notice of one younger. I know the tone so prevalent which either in jest or earnest ascribes any motives but the right one to a fellow in my position supposing I had, as I often wished, made some effort at becoming acquainted with you.

It is only a few short hours that I have to spend at Eton now, and grievous to me the thought is, for with all the wickedness I love the old place...I can only add that should we ever meet as I hope we shall in days when I shall be free from all these considerations, I shall hope to know you well enough to warrant my asking for a photograph, without needing an apology.[17]

This long and elaborate letter provides a peep into the steamy intimacies in Victorian schoolrooms. Edward Lyttelton rose to become Headmaster of Eton, Brodrick to be Secretary of State for India during Curzon's viceroyalty. Both always remembered the physical impact Curzon had made on them as young boys.

Over and over again, throughout his life the mention of Curzon's physical attributes appears like a leitmotiv. On his twenty-first birthday another friend, Richard Farrer, warns: 'Only beware of the besetting danger of any young man possessed of talent, position and good looks.'[18] Oscar Wilde, who was to befriend Curzon at Oxford, called Curzon, 'you brilliant young Coningsby'[19]and said to him he could 'never tire of hearing you being called perfect'.[20] Walter Pater, of the movement of art for art's sake, took him out to dinner more than once.[21] Edward Lyttelton after having once written a lengthy screed said, 'I must stop alas—what miserable things letters are: the writing of one only brings your rosy cheeks before me in a shadowy kind of way.'[22]

Another Eton Master called Ainger made seductive overtures to Curzon, inviting him on a holiday in London. Dod had immediately sat down to write to Lord Scarsdale a letter heavy with insinuation: 'I should be very sorry for any of my pupils to become intimate with Mr Ainger—of course it is a difficult thing to prove a man's motive.'[23] This time Lord Scarsdale forbade the visit. Ainger had offered to take Curzon for a holiday in London without first taking his house-master's permission. Curzon grumbled at not being allowed to go: 'I imagine that my tutor wrote and dissuaded you for giving me permission . . . I cannot see what Tutor has found to complain in it, unless it be his foolish jealousy at seeing any other master having anything to do with his pupils, which was the primary cause of the disturbance about Browning a year or two ago.'[24] But Lord Scarsdale proved surprisingly adamant.

To return to 'the disturbance about Browning', as Curzon put it. Browning had drawn young Curzon to his fold sending him the Prince Consort papers at the end of 1872. In the summer term of that year Curzon having been struck in the eye by a cricket ball found himself spending considerable time with Browning. Wolley Dod had complained to Browning:

> I strongly object to your taking Curzon out for drives without taking leave from me or the Head Master, to your writing to him by post, which you have done several times, when he is only two doors off, and most of all to your doing his verses for him, as I have suspected several times, and as he admits in the case of his iambics this week. I think the whole case is one which justifies an appeal to the Head Master, and I have accordingly made one, specifying the points on which your dealings with Curzon seem to me objectionable.[25]

Dr Hornby, taking up the cudgels for Dod wrote in a crushingly censorious tone,

> I want only to say first that in speaking of Curzon as an attractive

boy I did not wish to impute any motives to you, only to point out that in a public school appearances must be taken into account, and that, independent of a tutor's expressed wish, there is good reason why such an intimacy as seems to have arisen between you and Curzon should not continue.

I think I ought to say that habit which if I am not mistaken you have formed of entering into very confidential talks with many boys (not your pupils) about the character and conduct of their schoolfellows, seems to me to be a very dangerous one and to do great harm without really effecting anything for what, I believe, you have as your object in such intercourse, the eradication or diminution of gross vice in the school.[26]

Nevertheless, the friendship between the master and his young pupil continued. In a somewhat surprising turn of events, Lord Scarsdale came to his son's rescue, taking it upon himself to write to Browning exonerating him of having anything but the purest motives in looking after his son.

Exceedingly regret this extremely unpleasant complaint from Mr Wolley Dod with reference to your conduct towards my son George. I am fully aware of your warm feelings and keen desire that he should grow up a manly, true and pure-minded lad, and though it is possible that your notice of him may have served to annoy his Tutor, I give you full credit for acting from the purest motives, & I do not wish the kindly relations between you & my boy to fall through.[27]

The school nevertheless forbade the two to either meet or communicate. On 20 July 1874 Curzon wrote to Browning,

I cannot say how distressed I am that I am prevented from seeing you, all through the unkind and ungentlemanly and obstinate conduct of my tutor, whom I detest the more I see him. But I must thank you with my whole heart for all the inestimable good you have done me, for you have always been open to me as the best of counsellors, and you have warned me against evil companionship.[28]

Dr Hornby had his revenge on Browning a year later, sacking him on a small technical matter. But he was not able to drive a dent into the friendship. Curzon took Browning to meet his first wife, telling her that he owed all that he was to Browning. The master was also the Viceroy's special guest at the Delhi Durbar of 1903.[29]

The relationship between the two waxed and waned with Browning stoking the fire when it burnt low. After the initial burst of anguish at being parted from a beloved tutor, the intensity of Curzon's feelings subsided. Not allowed to meet, they had corresponded. Browning sent presents, wrote letters in violet ink and invited Curzon to Italy. Curzon replied to the Master affectionately but at fitful intervals, most of the letters are apologetic at not having replied promptly. On 6 August 1874 Curzon wrote woundingly from Kedleston, 'I have been so oppressed by the cares of a bachelor's life that I have not been able to find time to write to you.'[30] Browning promptly sent off two long letters and a volume of Tennyson.[31]

The new year began with yet another apology from Curzon, 'I am quite ashamed of myself for not answering your nice letter before.'[32] In May, Curzon thanked Browning for 'the very pretty pin you sent me by Graham' and said, 'I am afraid I forgot to thank you for your nice letter from London . . . I have still another thing to thank you for viz. the last number of the Tennyson.'[33]

In February 1876 Curzon wrote from Eton apologizing again for the delay in not having written earlier, saying, 'I have indeed been meaning to write to you for some time past but I am afraid I very frequently don't carry out my good intentions.' The letter, however, ends on a warmer note. 'I do miss you so much here, though I did not see very much of you for the last year or so, did I? Yet the place seems strange without you. I often wish you were back again if only for the pleasure of *merely* seeing you.'[34]

A year after Browning's departure Wolley Dod was reporting, with what seems like a sigh of relief, about Curzon to Lord Scarsdale: 'He has improved very much during the school times not only in work but in general conduct, as I hoped he would when older: his character, is becoming more manly.'[35]

Curzon, however, had not given up his friend. In the winter

of 1876, he asked of his father: 'The holidays begin on Friday the 15th. Do you object to my going from here to Cambridge to stay with Mr Browning who now lives there till Monday ... He is very anxious that I should go.'[36] Lord Scarsdale not merely allowed Curzon to visit Browning but also wrote to thank the tutor: 'My son much enjoyed his visit at Cambridge.'[37]

The indulgent Lord Scarsdale even sanctioned Curzon a trip to Europe with Oscar Browning. He also wrote to the Master saying he was, '...anxious for a chat with you as to the projected trip abroad with George after Christmas, which I am quite disposed to sanction.'[38]

Upon their return, Lord Scarsdale penned a letter of thanks 'to express my gratitude for all your kindness and attention to George during your recent trip. He tells me how much he appreciated your unvarying good nature towards him. I quite believe that your influence will have benefited him materially. I *expect* him to steer clear of all vice and contamination at Oxford and I have little fear of it ...'[39]

Eight years later, when Lord Scarsdale's younger son Frank applied to go to Europe with Browning, the father was not so forthcoming. While giving sanction, he grumbled about the costs. With eldest son George no such consideration had stood in the way. Curzon being his father's favourite, there was little that Lord Scarsdale would deny him. Of Frank, Lord Scarsdale wrote grudgingly to Browning, 'I don't in the least object to his going and am pleased he should have a friend of such experience as yourself as his *chaperon* ... but, as you know these outings cost money & youngmen don't always consider that.'[40]

Curzon continued to be the apple of his father's eye. Lord Scarsdale even corresponded with Browning about Curzon's university education, calling it, 'a serious and difficult question'. He proudly informed the tutor, 'you rejoice at his recent success—P. C. Italian prize! it did not altogether surprise me, knowing his capacity.'[41]

Upon return from his European tour, Curzon sat down to write a long letter to Browning:

We were very great friends before our tour—but if I may say so, we are still greater now; and the absence from your society short though it has been has shown me how much I value it. It was a cruel contrast exchanging your genial confidence for the dissatisfied and suspecting reception accorded to me by Stone [his new tutor]: and I am afraid I don't like the man. The masters here are full of inquisitiveness about our tour and beset me with questions.[42]

Thus the relationship of Curzon with Browning is seen to be characterized by intense but fitful feeling. There are the definite overtones of passion in Curzon's anguish at the Headmaster's decree that the two must not meet. Curzon had then ranted at Hornby: 'he cannot know how much he is doing by separating you from me.'[43]

Curzon also wrote the following to Browning:

I never go abroad without thinking of your pleasant experience in the past. Least of all I ever go to a place which we have jointly visited without lamenting that I have not the same admirable companion. I thought this a fortnight ago as I stood before the horn at lunch and paced the wooden bridge. I need no sooner did I arrive here—it was night—that I walked out to that self-same bridge for no other reason than that you and I did it in 1881, and that I wanted to revive the recollection.[44]

Significantly, the 20 year-old Curzon expressed a desire to make acquaintance with Lord Houghton. 'I should like to know Lord Houghton very much,' he said to Oscar Browning, 'you must introduce me.'[45] Richard Moncton Miles, Lord Houghton, politician and man of letters, had risen to fame in homosexual circles for his fine collection of erotic books and his friendship with the painter Simeon Solomon. Three months after his initial request, Curzon was reporting to Browning, 'I went to lunch with Lord Houghton 2 days ago.'[46] The friendship ripened quickly. Some time later Curzon said, 'Lord Houghton was kind to me and I saw him frequently.'[47]

But before too many conclusions are drawn about such things it needs to be remembered that most boys come from sheltered backgrounds, entering public school in a fairly innocent and ignorant state. However tempted they were to yield to seduction, out of curiosity or undefined feeling, much of the thrill must also have derived from the knowledge that what they sought was forbidden fruit. The secrecy and risk must have added to the excitement, the feeling being heightened by the distance separating the lovers. There was also the certainty, of course, of the impossibility of sexual fulfilment. Freud has recounted, 'I have never carried through any psychoanalysis of a man or a woman without discovering a very significant homosexual tendency.'[48] Though such schoolboy passions frequently arise in early adolescence in single sex schools, in most cases, as the boy leaves school the sexual impulse turns to the opposite sex.

Whether Browning made physical overtures to Curzon and whether they were successful is something that is not easy to establish. Browning confesses to having been very impressed by Curzon. Curzon too must have found some satisfaction in the relationship, otherwise the strong attachment between the two could never have been established or sustained. Transcending the relationship between a teacher and pupil, it was more of a friendship between kindred souls and Lord Scarsdale saw no harm in allowing this to continue. Curzon himself, strangely enough, does not seem to have suspected the master of having homosexual leanings, having once fondly said to him, 'you would have plunged headlong in . . . the Sea of Galilee and come out pure and sweet.'[49] On another occasion when Browning was on vacation, Curzon inquired with playful daring, 'are any of the same young ladies at your hostel . . . Do you dance?'[50]

This is not to say that Curzon was unaware of overt homosexual practice. Once on a trip to Europe he came across an instance and warned his friend Farrer:

A few remarks about Greece before I came to Egypt. Don't give anyone else recommendations to Hadgi Lazare. He turns out to be a g-gg-a. Thanks heaven I don't say so from personal experience though he did once look into my bedroom. But I found it at Corfu where he is well known and where I heard all about it from Admiral Baird.[51]

Nor was he unaware of the dangers of such a friendship. He wrote to the same friend about ' . . . going to a dance every night (excluding Saturdays and Sundays) but one.' He added:

But it is not so much to them that I point—as the principal sources of enjoyment as to the opportunities which seem to multiply year after year of strengthening old friendship and acquiring new . . . some of these insensibly develop into real affection: real but not dangerous.[52]

The last word is crucial for it shows Curzon's awareness of the hazards of such friendship.

Curzon nearly found himself engaged to be married when he was only 20 years old and still at Oxford. He confided to Brodrick about his narrow escape and Brodrick wrote back saying, 'You have more to beware than most from the danger of such entanglements.'[53] He found female company enchanting, women readily reciprocating to this handsome young man. 'At Perth, the women have figures that make one stare and itch,' he wrote to a friend while on one of his many tours.[54]

Twelve years after the controversy at Eton, when Curzon had entered Parliament, Browning was moved to write to Headmaster Hornby, going once more over the events that caused him, as he said, 'some of the acutest pain'. Hornby's withering reply was, 'I am sorry to say that your letter does not mend matters. George Curzon is, I think, a much stronger man than you imagine.'[55] Even the suspicious and sanctimonious Hornby had not suspected Curzon of having fallen prey to the snares of Browning.

Curzon seemed to have been attracted to Browning because 'he is the pleasantest of companions admirably informed on all subjects.'[56] Curzon was no doubt flattered because Browning took every possible opportunity to show how much he liked Curzon. Old friend Cecil Spring-Rice wrote to Curzon from

Balliol, 'O. B. was here for the Palmerston dinner and sat by me and talked about you.'[57] Curzon's vanity, his greatest weakness, needed to be constantly fed. Every word and gesture from Browning told him that he was a superior person. The friendship cost Browning his job. What greater compliment could be paid to Curzon?

Chapter 6
Oxford and Politics

Curzon went up to Balliol College, Oxford in 1878 when he was 19 years old. Old friends had already begun to refer to his tenure there as brief interval which must intervene between Eton and the Cabinet.[1] Within a short while, Curzon was to become the acknowledged leader of the Young Conservatives at Oxford. He soon made his mark at the Union and once again saw himself knocking off honours as he had done at Eton.

Lord Scarsdale had personally escorted his son to Oxford for the entrance examination. Before leaving Derby he had written to Oscar Browning, 'I intend going down to Eton, for George has to be in Oxford on Thursday the 22nd for his Exam. at Balliol.'[2]

St. John Brodrick, already settled at Balliol, welcomed Curzon, sympathizing with him over his spinal trouble. 'I was grieved to hear of all you have been going through physically with a most troublesome weakness and with actually so terrible an upset of all your plans and ambitions.'[3] The ever-admiring Brodrick even communicated with the Scarsdale family, entertaining them to dinner in London. 'I have met your people two or three times', Brodrick wrote from London 'and they dined with us one night this week. I was very pleased to make your sister's acquaintance and was struck by a variety of family traits.'[4]

From 1870 Balliol was stamped by the imprint of its brilliant Master, the Revd Dr Benjamin Jowett. For seventeen successive years, from 1888 to 1905, all the Viceroys of India were Balliol men. Between 1878 and 1914 more than 200 men joined the Indian Civil Service.[5] Balliol under Jowett exercized a profound influence on national affairs. It produced more scholars, philosophers and statesmen than any other college in the university. Jowett, always immaculately turned out in a tailcoat, with cherubic face and high-pitched voice, was a familiar part of the Oxford scene: Mathew Arnold, Tennyson, Swinburne were among the frequent visitors to his home. A popular jingle about him went as follows:

> First come I. My name is J-W-TT.
> There is no knowledge but I know it.
> I am Master of this College,
> What I don't know, isn't knowledge.[6]

Jowett's essay on *The Interpretation of Scripture* which came out in 1860 had increased the cry of heterodoxy against him. For years people thought he was a great heretic presiding over a college of infidels. 'His crime lay in saying that the Bible should be criticised like other books,' said his young friend and admirer Margot Asquith. She added how in his introduction to the *Republic* of Plato, Jowett had said:

> A Greek in the age of Plato attached no importance to the question whether his religion was a historical fact. Men only began to suspect that the narrations of Homer and Hesiod were fictitious when they recognised them to be immoral. And in all religions the consideration of their morality comes first, afterwards the truth of the documents in which they were recorded, or of the events, natural or supernatural, which are told of them. But in modern times, and in the Protestant countries perhaps more than Catholic, we have been too much inclined to identify the historical with the moral; and some have refused to believe in religion at all unless a superhuman accuracy was discerned in every part of the record.[7]

By the middle of the nineteenth century a great many intelligent people had begun questioning the traditional theology of the day. Darwin's *Origin of the Species* in 1859 and his *Descent of Man* in 1871 having struck a hard blow to the central biblical belief that man was a special creation wholly different from the animals. But while faith in Christianity waned, the vacuum, for some, was filled by faith in the Empire. In this transition the hand of Jowett is discernible. The highest ideal for a young man, according to the Master, was to direct boyish aspirations to the training for public service for the country and the Empire.

Nevertheless, the hold of the Church, by that strange Victorian paradox, survived openly avowed disbelief. Even Jowett, for all his intellectual heresy, did not give up his holy orders and his last words a few days before his death to Margot were, 'My dear child, you must believe in God in spite of what the clergy tell you.'[8] Similarly, Curzon, though he seems to have lost faith in Christianity, never renounced religion.

Though Jowett came from an impoverished background, he did not question the Victorian class division of society: the duty of the one being to govern and the other being to obey. For all his efforts to bring students into the same social scale, at heart, Jowett had a deep reverence for the fine manner, high tone, wide education and lofty example of the British aristocracy. When challenged about his partiality, Jowett said, '...one must remember how important it is to influence towards good those who are going to have an influence over hundreds of thousands of other lives.'[9] Jowett's faith in the lofty example of the British aristocracy found a ready echo in Curzon.

Generous, hospitable, encouraging with both sarcasm and praise, Jowett had singled Curzon out for attention. Along with Milner, Curzon was to rise to be numbered amongst Balliol's most prominent proconsuls.

Jowett remained a bachelor all his life. He enjoyed a long and sustained friendship with Florence Nightingale but, as his biographer points out, Jowett was separated by 'temperament and physique from the intimacies and habits of ordinary men.'[10]

Sensing in Curzon a promising pupil, Jowett had followed his progress with interest and on occasions had not hesitated to tender painful advice. Of Curzon's elaborate diction, Jowett had told him, 'I think you have many advantages and one disadvantage "Too much to say" in speech or in conversation. It is a good fault if *corrected* but a most serious one if left incorrected because it destroys the impressions of weight and of thought and gives the impression, probably very undeserved, of conceit and self-sufficiency.'[11] Nevertheless, Curzon could not break free from a prolixity of speech. Words rolled from his tongue like great waves of an ocean. They sounded well on the stage and in an Oxford Union debate or in Parliament but the stately patrician manner became an anachronism when carried into the twentieth century. This archaic and affected manner earned him much ridicule, but having set himself up in the style of a great aristocrat, Curzon would not change his ways.

The cloud of glory from Eton continued to blaze at Oxford. Curzon's striking appearance and self-confident, if not arrogant manner, coupled with his distinguished Eton record gave him a special position among his contemporaries and he received his full quota of adulation at Oxford. Not yet branded as a homosexual, Oscar Wilde became one of his many admirers. Attracted by his manner and looks, an American undergraduate wrote to him requesting a photograph. By way of explanation, the American wrote,

> As an undergraduate at Oxford I attended the Union dabates and noted the best speakers. As an unprejudiced observer I was interested more in studying the types of men than in weighing their political arguments. You were the only man I found who perfectly filled my ideal of what a young representative of the Conservative, and especially the aristocratic party should be. It was the intense aristocratic turn of your disposition which forcibly struck me; for which, indeed I had been abundantly prepared by works of fiction, but which I had never seen exemplified . . . You were to me a type.[12]

Like most young men of his background, Curzon's university education did little to equip him for earning a livelihood.

He read Classics which were divided into parts—Mods and Greats. He was required to spend six terms on each. For Mods he had to read Greek and Latin: Greats included History and Philosophy. Richard Farrer took it upon himself to warn Curzon to concentrate on his studies. He said, 'You are still young in Oxford life, and *do* devote the best of your faculties during the remainder of it to securing University honours: they are worth a great deal more hereafter than one has any idea of it at the time. I speak with feeling as having once underrated them and now regretting the mistake. They are not to be knocked off promiscuously like Eton prizes.'[13] It was good advice but Curzon was not inclined to follow it. Though in the Mods Curzon disproved Farrer's fear by taking a first, he was not so confident with the Greats, aware as he was that he had spent disproportionate time in the Union and the Canning Club. As the time of reckoning drew closer, Curzon applied himself to study hoping that his unorthodox methods, a hangover from Eton, would see him through. He confessed to Farrer, 'My history is pulling me up, backed by my translations. My logic and philosophy are remorselessly pulling me down. Which will win? Honestly, I think either. I tell you the truth. I do not think I am absolutely out of the chance of a first; but, upon my word, I am far from it...bearing in mind the superior importance attached to philosophy . . . I am inclined to believe that the betting is in on a second.'[14]

Nevertheless, he was considerably pained to discover when the results were out that he had not got a first. His friends were quick to try and console him. Alfred Lyttelton wrote, 'Of course you could have got the first class for certain had you denied yourself the Union, the Canning, and those other literary, social and political enterprises which have earned you the name of the most famous Oxonian in my knowledge of Oxford that I can remember.'[15] But Curzon had hoped to excel at both. Edward Lyttelton was also to reassure him: 'No one has ever got a first class and given himself up to so many other things as you and the real thing to think of is that you have got more out of the University than 99 out of 100 first class men, and that will not make the slightest difference in your numerous and trusting friends.'[16] His father gloomily wrote, 'I had a presentiment all along that you would not get that we wished for; although with very many a second would be highly thought of, I am quite aware that it has little value in your eyes: for there is, no doubt, in the estimation of the public, an enormous difference between the two distinctions.'[17]

While Curzon was still at Oxford the damaging 'superior person' verse had been put out about him. As an undergraduate he seemed to give the impression of a self-contained, almost precocious individual with fully-developed ideas, opinions and habits. His room at Balliol, like that of Eton, was characterized by a degree of lavishness which added to the impression of smug self-satisfaction. A contemporary wrote, 'They were furnished rather more elaborately than was usual; they were always kept spick and span, and gave a sort of impression of opulence and of that "superiority" which became crystallized in a famous phrase.'[18]

Jolted by his results upon finding that he was not as superior as he had begun to believe himself to be, Curzon felt something had to be done to vindicate his position. That summer he was to go on a tour of Greece, Egypt and Palestine. He decided to use his nights working for an essay towards the coveted Lothian Prize for History, Oxford's highest academic award for an undergraduate. He worked in the utmost secrecy, without the knowledge of his friends. He could not bear the mortification of another failure. The effort paid off and the prize did fall to him. What afforded him even greater pleasure was being once again able to prove to his friends that he could succeed in whatever he chose to do and in his own way. As he put it to his friend Farrer,

> You will perhaps have seen in papers or have heard from some card that I was lucky enough to pull off the Lothian and if so you will perhaps have wondered why you have neither now nor before heard anything about it from me. The fact is that I kept it universally dark. It would have distressed me to publish to the world another rebuff... and no one knew anything about it except the Brodder

[Brodrick] who detected my reading at the Brit Mus [British Museum] before starting abroad, and Edward and Welldon who often saw me writing while with them.[19]

It was this same desire that prompted him to try for another prize in History—the Arnold Essay on Sir Thomas More. The desire was whetted by the sight of a Balliol contemporary, Anthony Hope Hawkins, a runner-up for the Lothian, taking notes at the Bodleian Library. Curzon assumed Hawkins had set himself up to win the Arnold Essay and thereby avenge his earlier setback. Curzon's competitive interests were once more aroused and he promptly entered for the essay. He admitted, 'the undertaking was no light one, for I knew next to nothing about Sir Thomas More.'[20]

Curzon entered for the prize in December 1883, the rules dictating that the essay be submitted by the midnight of 31 March 1884. For the next three months he worked like a fiend. He records:

My day was spent as following: I rose at ten in the morning and breakfasted before 11. At 11.15 to 11.30 I was in my seat in the Reading Room of the British Museum. At 2.30 I took half an hour's interval for lunch and went to a stroll...At 5.30 fifteen minutes were allowed for a cup of tea.

He left the Museum at 8 p.m. but by ten o'clock he was home again and back '...in my chair from which I did not rise again until 4.30 or 5.30 in the morning.'[21]

He completed his essay on 31 March, and in a deliberately flamboyant gesture took the evening train to Oxford. Just as the clock was striking the midnight hour, Curzon woke up the custodian and handed him the essay. Curzon says, 'I apologised for my intrusion on the grounds that I was incommoding him in the interests of the prize winning essay.'[22] His audacity won out. Curzon did get the prize. Once again he felt he had disproved the verdict of his examiners who had not thought him good enough for a first.

Curzon had been elected Fellow of All Souls in the autumn of 1883, an honour which should have been sufficient to erase any pain left by his failure to get a first. Yet, still he could not let go any opportunity to prove himself. In the meantime, his platform oratory had won him attention in Tory circles. In March 1884, within days of winning the Arnold Prize, Curzon had been adopted as Conservative candidate for South Derbyshire. Edward Lyttelton had slyly said, 'I scent in your letter that you are making play not without success, among the celebrities of the world, notably those of your own party—it is a capital thing to be doing, and no better field can be chosen than Hatfield.'[23] Curzon's efforts paid off. The Prime Minister invited him to become one of his assistant private secretaries.[24]

While at Eton Curzon had begun to take an interest in politics. At Oxford politics had become Curzon's prime passion. Certainly, politically, the country was going through exciting times. Early in 1880 the general election had swept the Liberals into power with Gladstone at their head. Though Curzon had a special corner in his heart for Gladstone as an Etonian, his conservative instincts were repelled by his liberalism. Curzon promptly moved a motion in the Union regretting the results of the general election. The *Cambridge Review* reported that a thousand people had crammed into the hall to hear the debate. It went on to add, 'the debate ended with a House tumultuously impatient to hear Mr Curzon's reply, which indeed, was well worth hearing. He carried his audience completely with him, and his motion by a considerable majority.'[25] A month later Curzon was elected President of the Union by 308 votes to 193.[26] Asquith and Milner had been among its earlier presidents. That same year Curzon took upon the secretaryship of the Canning Club, the high bench for young Oxford conservatism. Neglecting his studies, Curzon threw himself wholeheartedly into bringing vigour into the club.

From the beginning Curzon was an ardent Tory, believing that the monarchy and the established church were the institutions that had raised England from a collection of petty principalities to a great power—and that they consequently had to be preserved and protected. An ardent follower of Disraeli, he believed that the world was divided into the classes and the masses, each having a predestined duty to perform as it were. He extended this same thinking to foreign affairs. As with men, certain nations were born to rule and others to be ruled. The British were naturally the chosen people but not any and every Englishman was automatically in a position to rule. Only those equipped by birth and education could do so. Having arrived at this conclusion, Curzon found it difficult to tolerate any-body holding views other than his own. While at Oxford he was writing,

> At a place and amid institutions whose roots are buried in the past, and whose history is intertwined with that of the nation, whose sons have carried its name to the corners of the world and stamped their own on the fabric of imperial grandeur, it would, indeed, be strange were there found any acquiescence in the sordid doctrines of self-effacement, in a policy of national or territorial disintegration, in the new-found obligation to shirk admitted duties, or in the application of the system of a parochial vestry to the policy of a colossal empire.[27]

The British political firmament was then ablaze with the tussle between Gladstone and Disraeli. Gladstone had entered Parliament on his first attempt at the age of 23. By the time Disraeli made his debut in 1837, Gladstone had served as Junior Lord of the Treasury and Under Secretary for the Colonies under Robert Peel. Tall, severely handsome and intensely religious, Gladstone could not have presented a starker contrast to the florid Disraeli. Gladstone's early inclination had been to join the Church, a wish he abandoned in deference to his father, the wealthy Sir John Gladstone who had made his fortune out of sugar plantations and the slave trade. Though he turned to politics, Gladstone never gave up his habit of soul-searching before arriving at a decision.

Disraeli, with his Jewish origins and raffish past, belonged to a different world. Unlike Gladstone, Robert Peel and John Russell who had been born to great wealth and privilege, Disraeli was born a member of the *petite bourgeoisie*. He started life writing novels and was often head over heels in debt. Disraeli had made it to Parliament on his fifth attempt only to have his maiden speech drowned out by derisive laughter. Forced to sit down, his speech incomplete, young Disraeli was said to have said, 'I sit down now, but time will come when you will hear me'. That Disraeli was able to break into the charmed circle and to rise to become not merely Prime Minister but also the closest confidant of the Queen is the stuff of which legends are made.

First becoming Prime Minister in 1868, Disraeli held power for a few months before the Liberals, under Gladstone, threw out the Conservatives. When he next came to power in 1874, and held office for six years, he dazzled his contemporaries, buying up the shares of the Suez Canal, and returning from the Berlin Congress bringing 'Peace with Honour'.

But despite the glittering flourishes, in the general election of 1880 the voters settled for the plodding Gladstone. Britain's population had risen from 27 million to 35 million while the liberalized franchise laws had widened the electorate. In India, Britain's image had suffered a severe setback in the abortive Afghan wars. Disraeli's choice of Lord Robert Lytton as Viceroy had not augured well. Shocking Calcutta society by flirting with pretty women, Lytton had embarrassed the home government by persisting with a disastrous policy in Afghanistan.

Nevertheless, Lytton served Disraeli's immediate purpose— in conjuring up a vision of Britain's great Empire by proclaiming Victoria Kaiser-e-Hind in a ceremonial that fired the imagination of the British. Dutifully, Lytton had convened on 1 January 1877 a truly imperial assemblage in the ancient Moghul capital of Delhi. Sixty-three ruling princes, arrayed in all their glory, had assembled to pay homage to their Empress, theatrically signifying, as Disraeli had wanted, their acceptance of the British monarch as their potentate. Reeling under famine and devastated by war, India did not take kindly to this expensive white man's ceremonial. Eight years later the Indian National Congress first met in Bombay.

While Curzon admired Gladstone as a man, he was worried by the Liberal PM's policy towards the colonies which seemed to him to be eroding the Empire. Gladstone had maintained with respect to India:

Let us only make common cause with her people, let them feel that we are there to give more than to receive; that their interests are not traversed and frustrated by selfish aims of ours; that if we are defending ourselves upon the line of the Hindoo Coosh, it is them and their interests that we are defending, even more, and far more than our own.[28]

When Disraeli died in 1880, Lord Salisbury, who had served twice at the India Office before succeeding to the Foreign Office, stepped into his place. Robert Arthur Talbot Gascoyne-Cecil, the 3rd Marquis of Salisbury was truly blue-blooded. Though his family was not as ancient as Curzon's, it was much more distinguished. His ancestor had served as First Minister to Elizabeth I. A hulking giant of a man with stooped shoulders, uncomfortable in society, Salisbury was happiest tinkering in his private laboratory at home. His countless experiments included those with electric lights and telephones and his daughter, Lady Gwendolen Cecil, recalled how visitors to Hatfield were once startled to hear Lord Salisbury testing a new hearing instrument by majestically orating 'Hey diddle diddle, the cat and the fiddle, the cow jumped over the moon'.[29]

Salisbury assigned to Curzon, his new Assistant Private Secretary, the Parliamentary seat of South Derbyshire, vacated by Lord Scarsdale's old friend Sir Henry Wilmot. However, his first attempt to enter Parliament in 1885 was a dismal failure. The new Reform Bill had extended the franchise in his constituency to another 7,000 voters, few of whom were prepared to vote for him because he was the son of a local lord. As he wryly reported to Brodrick, 'My electorate is 11,500, over 7,000 new voters. Of these between 4,000 and 5,000 are colliers and manufacturers (factory workers) and I haven't even a chance with them. They won't even hear me.'[30] He had tried to give them high-flown oratory in his best Oxford Union style.[31]

He never understood his electorate, nor did he try to. On the contrary he viewed them with near contempt. Though calling himself a politician, he never understood how to motivate men. Nor did he try to, relying upon his brilliance and industry to put him ahead of others. His aristocratic lineage and his education at Eton and Oxford had further sealed off a nature which took little interest in individuals as human beings. He remained conservative, orthodox, brimming with a sense of public responsibility suited to running the British Empire—but in an earlier era when it was not threatened by competition and technology. Vansittart, his deputy at the Foreign Office, was to say of him, 'He combined great knowledge with his innocence'.[32]

At his first election campaign in Derby Curzon had promised 'to maintain the integrity of the Empire'.

> The melancholy experience of the past five years with its terrible record of national humiliation, of squandered treasure and of wasted blood—offers a stronger condemnation of liberal policy than can any written or spoken words. The events which have occurred in South Africa, in Egypt and in Central Asia, and which have brought sorrow and shame to many a heart must have convinced the people of England that their interests are not safe in the hands of a Liberal Administration.[33]

The reception of his speech had been chilly. But characteristically he did not stop to ask himself why. He merely wrote to a friend, 'I wept less for myself than for the ignorance and backwardness of the voter'.[34] Fixed in his beliefs, he doggedly refused to look the future, not accepting that the masses were more interested in the problems of their daily lives than in the future of the Empire. He could never have remotely envisaged the reversal in his own lifetime of the conventional relationship of the classes and masses. Had anybody told him in the 1890s that he would live to see a Labour PM at 10 Downing Street, he would have dismissed the idea with contempt. In India too, he was to make the same mistake and refuse to see the writing on the wall.

Nevertheless, in his second electioneering attempt from Southport in Lancashire in 1886, Curzon was both much more direct and spoke with greater regard for the voters. In strong, forceful but specific rhetoric, he picked the theme of home rule for Ireland and its consequences for daily life. Though he began by denouncing the Liberal Prime Minister's Home Rule

Bill as being 'the latest attack on the integrity of the Empire', he moved on to what his voters wanted to know, saying that 'the consequent drain of capital from Ireland would drive many of her own sons in exile from the soil, and draft them as competitors into the already overcrowded labour market of England. To the British working classes it would therefore mean, on the one hand increased taxation, on the other diminished wages.'[35] This time he succeeded in getting into Parliament. He was 27 years old.

Chapter 7
Tales of Travel

Immediately after having made his début in Parliament, Curzon embarked on his first world tour. Travelling through Canada and the United States, Curzon visited India for the first time in December 1887. He returned to England the next year, only to leave that same autumn for a journey to Central Asia, which resulted in the publication of his *Russia in Central Asia*. Next year he was off to Persia and in 1892 published his mammoth 1,300 page treatise, *Persia and the Persian Question*. In 1892 Curzon embarked on his second journey around the world and subsequently produced *Problems of the Far East* which met with immediate success. In 1894 Curzon was off again to the Pamirs and Afghanistan.

With his characteristic fastidiousness, prior to embarking on his long journey Curzon made elaborate lists of the things required. His baggage included items as varied as insect powder and Eno's fruit salt to theatrical costumes and opera glasses.[1] During the trip he made a list of all the hotels he stayed at: the parsimonious habits acquired at Kedleston and encouraged at Wixenford made him keep a detailed record of distances and expenditure. He was able to work out the exact difference in expenditue incurred during his two journeys around the world. The first journey undertaken in 1887 cost him an average of £1 15s. a day, while the second undertaken in 1892, almost five years later, cost 2d. less.[2] This speaks volumes for Curzon's thrifty and careful habits.

By and large Curzon's hosts found him irresistible. He was not merely high-born, but had the correct aristocratic turn of disposition which had so fascinated his American contemporary at Oxford. Besides, Curzon was not averse to practising his own little deceptions if they afforded him entry into places normally forbidden. He thus managed to enjoy the hospitality of a wide variety of potentates, such as the King of Korea, the Emperor of Annam, the Mehtar of Chitral, the Shah Nasr-ed-Din of Persia, the King of Cambodia, the Amir of Bokhara and the Amir of Afghanistan. His wits stood him in good stead. In Korea, Curzon realized that he was not going to make much headway unless he could produce royal connections. A minister there had asked him if he was related to the Queen of England. 'No I am not', the irrepressible Curzon had replied, slyly adding, 'I am, however, as yet an unmarried man'.[3]

Nor did he hesitate to throw his weight about. When he embarked upon his second journey around the world in 1892, Curzon had already put in some years as Parliamentary Under-Secretary. His friend and companion Cecil Spring-Rice says there was once a shortage of ponies which delayed Curzon's arrival in Seoul. 'Curzon got very angry, explained that he was one of the most important people in England, and it was a matter of most vital importance that he should see the King next week; and he threatened beating and dismissals all around.'[4]

Similarly, he bluffed the Amir of Afghanistan into extending him hospitality for two weeks.[5]

Behind this façade of a clever, charming English lord, Curzon was passionately dedicated to the imperial idea. His travels were a part of an exercise to study at first hand the countries he knew he would one day be called to govern. His friends in Parliament took part in debates, made speeches and did the gilded round of Britain's great country houses. Curzon preferred taking arduous and sometimes hazardous journeys to far-flung outposts of the Empire. His travels filled him with national pride. The sight of Hong Kong, stirred him: 'No Englishman can land in Hong Kong without feeling a thrill of pride for his nationality. Here is the furthermost link in that chain of fortresses which from Spain to China girdles half of the globe,' he recorded in his first journey round the world in 1887.[6] The sight of Calcutta filled him not only with national but also family pride: 'Calcutta is a great European Capital planted in the East. The sight of these successive metropolises of England and the British Empire in foreign parts is one of the proudest experiences of travel.'[7] He also saw Government House, which as noted earlier, Curzon was to tell the townfolk of Derby on the eve of his departure to India as Viceroy, that its resemblance to Kedleston made him feel that India was beckoning him.

Curzon's increasing deification of imperialism, which was one of the driving forces behind his wanderlust, owes something to his loss of faith in Christianity. This is not to say that Curzon at any time of his life abandoned the Christian faith, only that he viewed it with increasing detachment. According to Lord Ronaldshay, Curzon's loss of faith began with his visit to Palestine in 1882. He says:

> It is uncertain when exactly full realisation of this intellectual revolt against the miraculous in the Christian doctrine flooded in upon him; but no one who reflects upon the self-examination to which he subjected himself as a result of his visit to the Holy Land ... can doubt in spite of his vehement protestation at the time, that it was then the corroding acid of scepticism first bit into his mind.[8]

On 1 March 1890 Edward Lyttelton wrote to ask, 'They tell me your church-going habits have broken down and that you have put away religion? Is this so? If so why?'[9] The Empire had become the new god in Curzon's pantheon. Nevertheless, Curzon remained a Christian. Prayer had been compulsory at Kedleston and Eton, and Oxford tolerated no lapse in church attendance. When Viceroy in India he was also known to go to church.[10] But it was the Empire which had become his true religion. Before his departure for India as Viceroy in 1898 he was to refer to 'the fascination and, if I may say so, the sacredness of India', with the reverence of a worshipper. He approached his work there as a mission.[11]

Seven years later, upon leaving India, his glorious viceroyalty in a shambles, Curzon's faith in the imperial mission remained undiminished. His farewell speech in Bombay had that same religious fervour as in the past.

Russophobia was no less responsible for Curzon's wanderings. Tsarist Russia had swallowed up Tashkent in 1865 and Samarkand in 1868. 'Are we justified in regarding with equanimity the advance of Russia across our Indian frontier?' a heated young Curzon had thundered in an Eton debate.[12] The visit to India made him obsessively aware of the danger to her frontiers posed by the Russian empire and it was to study that threat at first hand that he undertook a journey to Central Asia. The resultant publication, the 400 page *Russia in Central Asia*, was to describe the 900-mile long recently opened Trans-Caspian Railway. The book's real intent was to assess the advent of the Russian giant in Central Asia. It was this obsession with Russia's designs upon India that made Curzon undertake the long, hazardous and uncomfortable journey from Merv and Bokhara to Samarkand and then to bump his way in a horse-driven carriage for the 190 miles to Tashkent. Even as a schoolboy he was aware that the Russian question could not be considered in isolation from European affairs. Believing this, as Viceroy he persisted in pursuing a 'forward' frontier policy for India which was to embarrass the home government.

The motive for his Persian travels which began in September 1889 was the same. For nearly six months he travelled alone, riding almost 2,000 miles on horseback. Once again, the essence of his interest was to gauge Russian expansion on Persia's northern borders. Curzon's own policy did not envisage the use of arms against Persia; only the provision of an economic and political blockade on the Russian sea-route to India. *Persia and the Persian Question* was published in 1892 after three years of labour. The two volumes amounted to 1,300 pages.

Soon Curzon was to set off for Afghanistan. Curzon was to write, of all the countries lying on the glacis of the Indian fortress, Persia, Baluchistan, Afghanistan, Tibet, China, 'I was beyond measure desirous to visit that one of their number which, though perhaps the most important, was also the least accessible, and to converse with the stormy and inscrutable figure who occupied the Afghan throne, and was a source of such incessant anxiety, suspicion, and even alarm to successive Governments of India as well as India Office in London.'[13] That Curzon should foresee the need for establishing personal rapport with such a crucial neighbour, however dangerously whimsical, was a tribute to his vision as a statesman. That he felt himself confident of being able to achieve it must be taken as an exercise in high courage.

Permission to make the trip had been slow in coming. The Viceroy, the 9th Earl of Elgin, was hesitant. Curzon said, 'The Home Government viewed my project with some anxiety; the attitude of the Government of India was veiled in chilly obscurity.'[14] Finally, when the grudging permission came, Curzon was told coldly, 'the Government of India would assume no responsibility for my safety.'[15]

A journey to Afghanistan was high adventure. The ferocious warlike tribes viewed the advent of a foreigner, even one promised safe conduct by the Amir, with dire suspicion. Besides, the Amir himself was wily and capricious, his cruelty legendary: for crimes like theft and rape he had the guilty blown from cannons or frozen alive.[16]

The route from Kashmir to Gilgit to the Hunza valley was bisected by glaciers, with eight mountain peaks of over 24,000 feet. With a single European companion he pressed on to the plateau of the Pamir at an elevation 14,000 feet. His purpose ostensibly was to find the source of the River Oxus and to cross the roof of the world, but even more, to gain reassurance for his ego, to prove that no physical feat was impossible for him.

There were glaciers, torrents, snowstorms to cross, he had to ride on razor thin paths cut in precipitous mountain ridges: one pony died of exhaustion, another disappeared down a steep ravine to instant death.[17]

The Amir, however, proved to be a warm and indulgent host. Enchanted with the costume devised by Curzon for his state entry to Kabul, with its theatrically gilded epaulettes and their ample bullion, the Amir had promptly backoned his court tailor. Pointing out the glittering appendages, the Amir impressed upon the man how they were of 'a character necessitating serious notice and even reproduction at the Court of Afghanistan'.[18] The Amir and Curzon remained life-long friends and when upon his marriage Curzon sent the Amir a photograph of Mary, the Amir wrote back warmly saying, 'I also congratulate you, my honest friend, that you have married one wife, she is competent... If she should at any time thrash you I am certain you will have done something to deserve it.'[19]

These travels were to Curzon a part of the white man's burden. The orientals could be shifty, 'shocking thieves and rascals' but their upliftment was a God-given mission for the British.[20] He travelled to gain knowledge, not to strike roots and make a fortune as many earlier British nabobs had done. Not that he was not in need of money—the allowance that Lord Scarsdale gave him, of £1,000 a year, was a pittance when compared to the income of most of his circle. The profits from his articles and books were merely sufficient to cover his travelling expenses.

His journeys also offered other opportunities. Curzon records how once during a tour his host, a Sultan, sent him a concubine. After dinner when he retired to his bedroom he found,

> ... an exquisite little creature waiting for me beside my couch sent there by my brother for my pleasure. It took all the strength of my character to send her on her way, and all my charm and tact to persuade her to go without hurting the sweet child's feelings.[21]

He sent her away but not before he had 'allowed a time to lapse before her departure, so that she would not be blamed for having failed to please me.'[22] His gesture reveals a surprising and touching concern.

Other sights also awaited him:

> Boy brothels were well-established in Naples, as well as in Cairo and Karachi; paederasty was endemic among the Persians, the Sikhs and the Pathans, and was elevated into an integral part of Afghanistan culture in the form of batsha troupes of dancing boys.[23]

Curzon seems to have been more attracted to white, English, upper-class women, however. From Cairo in 1883 he wrote to his confidant Richard Farrer,

> The most charming person I met there was the Baroness de Malortie (English herself but married to a foreigner) whose face, an extremely beautiful one, is familiar from photographs in Bond Street windows. I got to know her most intimately (this for your private ears only) and I count her one of my dearest friends.[24]

During his travels Curzon saw what he wanted to see, namely that Great Britain was the greatest instrument for good the world had ever seen. 'The peasants are devoid of truth and all its attributes—candour, frankness and honesty. They are treacherous and deceitful,' he wrote in one of his journals.[25] Airing his contempt for those people who had not the good fortune of being born British, he dismissed the deposed King of Oudh because he was 'a true son of the East . . . who devoted his enormous wealth to the construction of a bizarre palace with innumerable pavilions and tanks and glass lamps and bad copies of European pictures of a pornographic type'. Curzon added patronizingly, 'The life of these oriental despots must have been a curious mixture of pomp and frivolity just as their minds were a rule about equally compounded of childishness and vainglory.'[26] The Bengalis he dismissed as 'not an inspiring or manly race'.[27]

The Japanese, however, met with his approval: 'They are neither such robbers nor quite such artful deceivers as their brethren in Cairo or Constantinople.' [28] In general, everyone except the British were liars and deceivers. These sentiments were largely a part of the social arrogance of the governing classes of Victorian England. Viewing with contempt anybody who was not of the British upper class their behaviour bordered on the boorish and uncivilized. Alfred Lyttelton once threw a half-sucked orange from a moving train at the face of an innocent Italian gentleman who happened to be standing on the platform in Venice merely to laugh at the man's discomfort.[29]

Smug and patronizing, Curzon freely passed judgement as he went. The Greeks, who had bequeathed the democratic idea to the West, were not suited to democracy because, 'A people just awakening from the night of four hundred years of Turkish oppression is hardly fit to receive the mead of full enlightenment.'[30] The Egyptians were not spared either: 'Civilization is foiled by a country which refuses to be civilized, which will remain uncivilized to the end.'[31] On a ship to Canada, he wrote witheringly of his fellow passengers: 'There are few ladies or gentlemen among them. The social status of the remainder is indicated by the aristocratic names they bear—Tulk, Tottle and Thistle.'[32]

Where the lower order of Englishman did not come in for contempt, was where he happened to be manning an imperial outpost. 'The same high tone exists through the various strata of society and employment, and the clerk behind the counter of the English bank will be no less a gentleman both in birth and education than the Governor in his palace or the Minister in his legation,' Curzon noted with satisfaction.[33]

When he saw the Taj Mahal for the first time he was entranced. For once the young imperialist was spell-bound. When he found his voice it was to say, 'there is a fascination and poetry' about it 'which are quite undescribable'. He humbly admitted that before seeing it he had thought it 'comparable to the Alhambra at Granada—the most beautiful Sar[acenic] building in Europe.' But having seen the Taj, he confessed 'I am aware no such comparisons are possible.'[34] Nevertheless, he could not help taking satisfaction in the British presence, for the garden at the Taj was looked after by a Britisher. 'It is difficult to exaggerate the extent to which the beauty of the Garden contributes to and enhances the Taj... It is to the credit of England that this garden is mainly the product of English hands, a burly Yorkshireman named Smith having been its custodian for some 20 years.' Which for Curzon went once again 'to prove the dominion of English idea.'[35]

Chapter 8

George and Mary

While still at Oxford, full of serene confidence, extraordinarily handsome, clever, much sought after, Curzon had romped through England's great country houses, talking politics through the night, conducting amorous liaisons. He even had the effrontery to play tennis stark naked before breakfast with friends from the all-male Crabbet Club.[1] It was at this time that the doggerel referred to earlier first surfaced.

Lord Ronaldshay says the original version was a parody of a song written by Curzon himself for *Waifs and Strays*, an Oxford poetry magazine. That parody had first seen print in the *Balliol Masque*, a collection of rhymes mainly by J. W. Mackail, later Professor of Ancient Literature to the Royal Academy, and H. C. Beeching, later Dean of Norwich. The version in the *Balliol Masque* went:

> I am a most superior person, Mary,
> My name is George N-th-n-l C-rz-n, Mary
> I'll make a speech on any political question of the day, Mary,
> Provided you'll not say me nay, Mary.[2]

The better-known version that has come down to posterity has been ascribed to Cecil Spring-Rice.

This tendency to make jokes at another's expense was to

rebound badly and Curzon was to become a martyr to his own ridicule. The jokes he made about himself were repeated as examples of his pompousness and conceit. Years later he was to regret the 'superior person' verse, bitterly believing that it had been responsible for all the great misunderstandings about him. In a frank talk with Lord Riddell towards the end of his political career, he said:

> I have always been misunderstood. It has been assumed that I am a pompous person, loving display and ceremony and devoid of any sense of humour. This is due, in great measure, to the well-known skit about George Nathaniel Curzon being '*a most superior person*'... It arose out of the fact that one night, owing to fog, I was compelled to stop at Blenheim. This led to a lot of good-humoured chaff... When I went to India, the skit followed me there... My reputation is due in some measure to the fact that for many years I have been braced up with a girdle to protect my weak back. This gives me a rigid appearance which furnishes point to the reputation of pomposity.[3]

A year and a half before his death Curzon wrote to his second wife, 'Mackenzie King the Canadian Prime Minister came in to say goodbye... He said he had come to the conference with a violent prejudice against me based on the newspaper pictures of the superior person... Oh! How these cursed papers have killed me...'[4]

But at the time he had done nothing to check the tide. A superior person was what he was and what he wanted to be acknowledged as. He saw no reason to abate his vanity or to conceal it. Reginald Brett had warned even while he was at Oxford: 'Dear boy... guard against flattery, if you will, of people who desire to please you.'[5] Curzon did not think it necessary to follow the homily. On the eve of his marriage in 1895, Brodrick, the most tenacious of his flatterers, was to write:

> You know how high I rated the promise of your life as well as our friendship from the moment you set foot in Oxford—But probably your best wishers hardly contemplated a more brilliant career than you have achieved. You have won the ear of the House of

Commons and of the country on certain subjects on which your reputation is unique; you are universally marked out for high office; you command great audiences.[6]

Eventually it was Brodrick who was instrumental in the humiliating termination of Curzon's viceroyalty.

In childhood and adolescence, however, practically every force surrounding Curzon had told him that in physical appearance and intellectual powers he was an excellent human being. How could he not but feel superior? 'Out of his large family,' wrote Irene, 'my father seems to have kept for himself all the gifts of statesmanship, erudition and rhetoric that made him what he was in later life; he left his brothers and sisters at the post at school and university.'[7]

By his early twenties Curzon had grown into an extraordinarily handsome young man. His face still had a boyish pink and white complexion even though it had lost its schoolboy roundness. His chiselled lips had the correct disdain. His face was now lean and sensitive. He could appear cold and haughty to strangers but he was capable of breaking into infectious laughter in congenial company. Alfred Lyttelton said longingly, 'I would very much enjoy the sight of your shapely figure.'[8] Another admirer was Margot Tennant who wrote flatteringly to Curzon after a visit: 'It was so delicious seeing you again after such ages and you were in such splendid form...you quite shook up the whole evening. Alfred said you gave us all a new lease of life.'[9]

At 23 Curzon found the doors of most of England's great houses opening to him. Curzon and his friends sailed along in this circle presided over by Prince Edward. But they had also formed another exclusive coterie of their own. Known as The Souls, it was made up of the ambitious, clever, well-read, witty, elegant, high-born, and more importantly, those who were aware they were so. The urbane Arthur Balfour, later to become Prime Minister, the Tennant sisters, daughters of the wealthy Sir Charles Tennant and St. John Brodrick, were among 'The Souls'.

The all-male Crabbet Club, founded by traveller, writer,

politician Wilfred Scawen Blunt and the centre of the fashionable, flippant intellectualism of the early 1800s also made much
of Curzon. Once a year, several rising young politicians met at
Crabbet, the country seat of Blunt in Sussex 'to play lawn tennis, the piano...and other instruments of gaiety. To write
bout rimes, sonnets and make sham orations...You will find
young Radicals and Tories, amateurs of poetry and manly
sports. The President at dinner in the costume of the Arab
Sheik...'[10]

Even in the 1880s, England continued to adhere to the old
hierarchy. 'In those days it [society] was still relatively small
and to be admitted to it required birth or introductions and
was still regarded as a privilege,'[11] remembered Curzon. Cau-
tiously, room was being made for the new industrial magnates
so attractive to the Prince of Wales. Victoria remained in retirement, a shadow of her old self, but Bertie had embarked on a
giddy round. His parties at Marlborough House went on till
the early hours of the morning as giggling guests took turns at
tobogganing down the stairs on huge silver salvers.[12]

Country house parties began with guests arriving on Tuesday,
but leaving on Saturday; the Sabbath still being a dreary day.
The great houses of London considered it their obligation to
give balls to the accompaniment of Hungarian and Viennese
bands with damask-covered tables groaning with delicacies.
Their expenses sometimes so effectively crippled the hosts that
they were never able to repeat their hospitality. Curzon recalls
going to a ball at a house in Grosvenor Square where the
freshly-cut hot-house flowers were specially brought in from
France, a fact which drove the Vicar of St Peter to deliver a
sermon on the folly of extravagance the following Sunday. A
clergyman's son, Curzon reveals, 'I was present at the ball and
[later] heard the sermon'.[13]

As the superior person verse suggested, Curzon was a
particularly eligible bachelor. Even before he had left Oxford,
more than one woman set her heart on him. He had captivated
all the Tennant sisters but was not prepared to get captured
himself. His brush with promiscuous male attention at Eton

had not soured his feelings for women and he enjoyed flirtations at country-house parties. And as already mentioned, in the spring of 1883 he had an affair with a baroness in Cairo. Attracted by and attractive to women, he was, however, reluctant to commit himself and in a privately circulated pamphlet of the Crabbet Club he put forth in humorous verse his views on marriage.

> Perchance with all these gifts
> you'll say, its strange I am not wedded,
> And preach a sermon on the woes
> of life when single-bedded,
> But if Clarisa I adore, and rashly
> go and marry her,
> To Chloe's subsequent embrace
> it may erect a barrier.[14]

The four Tennant sisters adored him and he gave something of himself to each one of them. Charty, married to Lord Ribbelsdale, wrote to him, 'you love several, but I feel proud to be amongst them.'[15] Margot was probably his most ardent admirer. Four days after her marriage to Henry Asquith, she wrote to Curzon:

> Dearest George, Your letter was a great delight to me and I shall keep it for my life. If I cd. [could] really think I had brought 'real good' to one who spreads everything dear that the world can give around him why I sd. [should] feel deep down pride but oh! George I'm not good I'm only living and *I love you*.[16]

In her autobiography published in 1920, the same Margot was to give a rather uncharitable description of her old friend:

> He had appearance more than looks, a keen, lively face and an expression of enamelled self-assurance. Like every young man of exceptional promise, he was called a prig...
>
> He had ambition and—what he claimed for himself in a brilliant description—'middle-class method': and he added a kindly feeling for other people a warm corner for himself... He was chronically industrious and self-sufficing; and even though oriental in his ideas of colour and ceremony, with a poor sense of proportion and a childish love of fine people, he was never self-indulgent. He

neither ate, drank nor smoked too much and left nothing to chance.[17]

Margot's uncharitable assessment may have owed something to her having been spurned by Curzon. Biographer Leonard Mosley claims that Margot married Henry Asquith on Curzon's recommendation. Not wishing to get entangled himself, Curzon had apparently told her that Asquith would give her 'devotion, strength, influence and a great position'.[18] Margot seems to have followed the advice and on acquaintance was reporting to Curzon about her future husband, 'Mr Asquith who improves with success, I think he is more flexible and can be delightful company.'[19]

More serious, however, was the attention Curzon had begun paying to the beautiful Lady Sibell Grosvenor. Four years older than him, Sibell had married the son and heir of the Duke of Westminster in 1874 but continued to have her bevy of admirers. Part of Sibell's attraction may have been the large fortune she was supposed to bring with her. While he was at Oxford, Curzon had begun paying her court.[20] But when widowed after ten year's marriage, Sibell chose for her second husband the soldier-poet George Wyndham. She and George Wyndham remained Curzon's life-long friends. Charty Ribbelsdale had rushed to console him:

> I am disappointed in Sibell. She has descended fathomless fathoms in my estimation.[21]

Curzon's self-confidence could not but bounce back in the face of such comfort. Curzon was determined not to marry until his career was established and then only to someone who would help him by way of money in furthering it.

In 1890 Curzon first met 'the dearest girl I have met for long. That girl is Mary Victoria.'[22] When he saw her he was 31 and she 20. She had beauty, money and charm; he birth, breeding and a brilliant political future. Both were ambitious. For her it was love at first sight. He was attracted to her: there was nobody better able to bolster George Nathaniel Curzon's confidence in himself than the exuberant Mary.

Her father was the Chicago millionaire Levi Leiter, who had vast stakes in real estate in the Midwest and was, in spite of his name, a convert from Dutch Calvinism to the Episcopalian Church. She was also known to be the heiress to quite a large part of her father's fortune.[23] The descendant of a blacksmith, Levi Leiter was not educated beyond high school. A fierce zest for money had driven him to leave his job as a shop-boy in Maryland to find his fortune in Chicago. Within his first decade there he had managed to buy, with his partner's help, the department store that had first employed him. After that there was no looking back. Cashing in on the great Chicago fires, he purchased charred sites and built on them at enormous profit.

Mary was his eldest daughter and second of his four children. But she was his favourite. Mrs Leiter, a school-teacher before her marriage, was an incurable social-climber. Her daughter's biographer says '...to most she appeared bold, proud, ignorant and selfish, ambitious to make herself a great social position.'[24] Every year she dragged her children to Europe to prepare them for the social peaks she hoped they would scale.

On their fourth visit to Europe the Leiters stopped in England, where the Duchess of Westminster invited them to her ball. Mary, with her entrancing eyes and youthful bounce, captivated the guest of honour, no less a person than the Prince of Wales. He walked over to her and invited her to open the ball by dancing the first quadrille with him.[25] Old British dowagers craned their necks to peer at this new American girl who had caught the Prince's eye. Curzon first saw Mary that evening.

Soon after they were both guests at a country house party at Ashridge. Here, within the splendours of graciously panelled rooms hung with Van Dycks and Rubens, the two met and talked. Mary was awed by the surroundings but even more by her companion.[26]

In London Mary found herself fêted. But the one man she cared most for could be very distant and Mary was puzzled and hurt. But he refused to be baited.[27] Once he did not answer her letter for months and when he did it was only to say woundingly, 'Now at last I have sat me down to work off a lot of arrears.'[28]

More torment was to come. Two years later on a trip around the world, Curzon was in Washington, D C. He passed the Leiter House in Dupont Circle and yet did not bother to call on Mary. Instead he went to Virginia and spent a couple of days at the house of a Mrs Rives, whose daughter Amelia had captivated him with 'the undivided insistence of her starlike eyes.' Curzon had exclaimed, 'Oh Lord the nights on the still lawn under the soft sky with my sweetheart.'[29]

But Mary refused to be vanquished. She had followed his travels with care and must have known of his visit to Washington, if not to Amelia Rives. Mary confessed to a friend, 'I will have him because I believe he needs me.'[30] She felt no shame in pursuing him and in doing so unconsciously helped to swell Curzon's ego. He could be cruel to her and yet she remained faithful.

Mary was in Paris with her mother in the spring of 1893 when she received news that Curzon was also there returning home from his second trip round the world. She had not seen him for eighteen months. He had been to Washington, D C and not bothered to contact her. Nevertheless, she was too overjoyed to bear any grudge. It was she who took the initiative and arranged a dinner meeting at the Hotel Vendome. It was there that Curzon at last proposed marriage.[31]

He had little intention of doing so when he came to dinner but had been flattered enough by the account of her adoration into doing so.[32] The question must arise as to why Curzon eventually settled on Mary—an American with a name like Leiter—as a marriage partner. She was beautiful and sophisticated, but the stigma of a Chicago ancestry could not be brushed aside. She was rich and Curzon surely needed money, but yet the marriage was not merely for money, for this 'superior' young aristocrat could have found another equally wealthy British debutante. The Tennant sisters had showed him in many ways where their affections lay. But it was Mary whom Curzon finally chose. Perhaps Mary's uncritical adoration won his heart. She played to his vanity, assuring him that in their relationship she would willingly accept him as her master.

The morning after the proposal he had sent a note around to Mary. Dated 4 March 1893, he said, 'you were sweet last night, Mary, and I do not think I deserved such consideration. While I ask you and while you consent to wait, you must trust me, Mary, wholly, even as I trust you, and all will be right in the end. I will not breathe a word to a human soul.'[33] Curzon seemed to want to have his cake and eat it too. Once again Mary obediently complied with Curzon's suggestion that the engagement should remain secret. She did not question the importance of his journey to Afghanistan. He by now knew the whole of the East except that important buffer state. She realized how vital the journey was to the career he had fashioned for himself. She realized also the heavy risk entailed and valiantly told him that if he died on his travels she would never marry but would instead try to release Kedleston from debt.[34]

Soon after these solemn promises, Mary got reports from London to say that Curzon continued to move through society as if he were a free and independent bachelor. A few weeks after his secret engagement, his old master Dr Jowett was asking when Curzon intended to exchange All Souls for one body![35] But Mary took her beloved's errant behaviour with equanimity. There seems little reason, however, for such callous behaviour on Curzon's part. He wanted perhaps to tantalize Mary. Perhaps his overweening vanity needed to be pampered by the knowledge that Mary would be prepared to wait no matter what he did. Audaciously, he wrote, 'I am spared all the anxiety of courtship, and I have merely, when the hour strikes, to enter into possession of my own.'[36] Such observations were hardly flattering to his betrothed, yet she did not protest.

Back in England at last in January 1895, Curzon journeyed to Kedleston to inform his father of the engagement. In spite of being engaged for almost two years, Curzon had not yet spoken of Mary to his father. He was then 36 years old and an under-secretary in the Government One biographer has explained Curzon's hesitation to broach the subject to his father as indicative of a defective parent–child relationship.[37] But what emerges is that because of Curzon's strong attachment to his father, it was very important to him that he approve of the match. Curzon had perhaps feared that his father might object to the Leiters because their name was of Jewish origin. But as he happily wrote to Mary ' . . . a sight of your lovely face on the smaller Miss Hughes photo (profile, white dress, hand behind back) completely convinced him'[38] The Leiters set their seal of approval upon the nuptials by making a marriage settlement of £ 140,000 for Mary, which upon her death would pass down to her children.[39] Levi Leiter also made out a yearly stipend of £ 6,000 for his son-in-law to enable them to move in the circles which would help with Curzon's political advancement. Curzon could provide only £ 1,000 a year, 'a paltry sum, I fear, but my family is poor'.[40]

The marriage date was fixed for 22 April 1895, the place as St John's Church, Washington, D C. But just when everything was going smoothly Mary began to feel qualms of uneasiness and she, for the first time, gave Curzon anxiety in their courtship. Cecil Spring-Rice, Curzon's old friend who was at the embassy in Washington and who had once been one of Mary's suitors, had told her of Curzon's philandering habits, adding that Englishmen do not set much store by marriage: 'A man chiefly has a home to stay out of it.'[41] Curzon, who was in bed at Kedleston with back trouble was much alarmed. He wrote:

> You must not let Springy [Spring-Rice] talk Nonsense to you, darling . . . he cannot forget that he loves you and he cannot get the better of his jealousy. I think we should be lenient, darling, I do not understand jealousy myself . . . Then, about your beauty (oh, that man is black with jealousy, that is it), it is going to be not merely my heart's treasure, my arms' possession, but my glory and my crown.[42]

It is interesting to note what Curzon says about jealousy. He did not understand jealousy himself, though may be this was because he was too egotistical to be jealous.

Mary's doubts persisted and Curzon for the first time in his dealings with Mary found that he was not really spared all the anxiety of what is called a courtship. He had to bring his persuasive charm to play. In bed with a bad back he collected himself to write:

> Sweet child you ask whether I am prepared for a somewhat demonstrative affection! Darling, I should be miserable if I did not get it.[43]

Curzon sailed to Washington, DC on 18 April. His brother, Frank, travelled with him, carrying the Kedleston diamonds for Mary to wear.[44] The Curzons were married on 22 April 1895 in Washington, D C after five years courtship.

When they returned to England, Lord Scarsdale rose to the occasion by receiving his new daughter-in-law with a ceremonial befitting a future mistress of Kedleston. The lavishness of the reception was in itself an indication of how deeply the father cared for his eldest son. Lord Scarsdale had received the newly-weds at the station in two great barouches and personally escorted them the two and a half miles to Kedleston to the accompanying welcome of church bells. Abandoning his customary frugal habits, he gave the new couple a fabulous feast. Five hundred and fifty tenants were invited and they gave the bridal couple a huge silver tray. A band stood in attendance while Curzon went around shaking hands and joking with the guests.[45]

The marriage that had begun as a seemingly calculated match, turned out to be one of the hauntingly beautiful love affairs in history. Unflawed till the end, the union was based on intensely passionate physical love. Mary gladly accepted her role as the subservient partner. As she once wrote proudly to her father, 'George will do with his career what he chooses and *nothing on earth* can alter his iron will. I have long since realised George's iron will and never crossed it.'[46]

Back in London, with the help of his father-in-laws's bounty, Curzon was able to take a lease of 4 Carlton Gardens which had been built in the Regency style by the architect John Nash.[47] Situated near Buckingham Palace, it was one of London's most aristocratic town-houses. Later the same year, the Curzons also took a place in the country, *The Priory* in Reigate, Surrey, a place for weekends and country house parties.[48] At long last Curzon was in possession of two beautiful homes, considerable wealth and a lovely, devoted wife. He was now in a position to live and entertain in the style he had always longed for and which he considered necessary for his political advancement.

In his characteristic manner, Curzon took over the management of the houses. The training under his father and under Dunbar at Wixenford had made him fussy and meticulous. He reported to Mary,

> The house [Carlton Gardens] looked charming today. The drawing-room pretty, the staircase (now quite finished) a little dark but decidedly handsome, the other rooms nice. I am going to have the colour of the outer hall slightly altered. All carpets and curtains to be in by Christmas.[49]

Getting good servants to come and work for him, however, was to become one of the more taxing problems of his life. Mary did not seem to mind her husband's interference in what was then strictly a woman's domain.

The summer had scarcely begun when the young Curzons moved into their new home situated in the heart of London's most aristocratic district. London's season was in full swing and would continue its merry-go-round until the end of July. The Prince of Wales entertained them at home: he had not forgotten the young debutante he had met at the Westminster Ball. But such outings were now few and far between; life was no longer measured by the rhythm of fun. As a debutante, Mary had whirled through a round of parties but now she was the wife of a junior minister who took his duties seriously. Lord Salisbury had elevated Curzon to Under-Secretary of State at the Foreign Office and also made him a Privy Councillor. Nevertheless, his new political position did not take Curzon far in the social hierarchy.

But on 11 August 1898 all this changed when the formal announcement of Curzon's appointment as Viceroy was made.

Mary's delight and triumph had known no bounds as she rushed off to tell her parents:

> It takes my breath away. For it is the greatest position in the English world next to the Queen and the Prime Minister, and it will be a satisfaction, I know to you and Mamma that your daughter Maria (*sic*) will fill the greatest place ever held by an American abroad.[50]

Mary was 28 years-old. Curzon promptly wrote to his father to share the good news with him: 'Lord Salisbury has offered me with the consent of the Queen the Viceroyalty of India and I have accepted it. I have told no one else but Mary.'[51]

The Viceroy-designate had previously held no high office. The Marquess of Lorne, a grandson-in-law of the Queen had been angling for this prize. Nevertheless, Curzon had written to his chief, Lord Salisbury, to press his claim:

> I have long, however, thought that were the post in India to fall vacant while I was still a young man—I shall be 40 by the end of Elgin's term—were it to be offered to me, I should like to accept it . . . I have for at least 10 years made a careful and earnest study of Indian problems, have been to the country four times, and am acquainted with and have the confidence of most of its leading men . . . I have been fortunate too in making the acquaintance of the rulers of the neighbouring states, Persia, Afghanistan, Siam, friendly relations with whom are a help to any Viceroy.[52]

Both Curzon and Mary were so overwhelmed by their new glory as to never quite recover their equilibrium. They were invited to Windsor, where the old Queen put on her large pair of glasses to inspect the young American girl whom destiny had catapulted to the position of her Vicereine. Satisfied with what she saw, Victoria told Curzon, 'I must congratulate you for your wife is both beautiful and wise.'[53]

A title was necessary for the Viceroy-designate for he was the personal representative of the Queen in India. Curzon was not very happy at the thought, for it meant giving up his position in the House of Commons on his return from India. Several times as a parliamentarian he had tried to find a solution to the custom that required a member to give up his seat on succeeding to a peerage.[54] Finally he chose an Irish title with the family name of Curzon since it did not qualify him for the House of Lords. When the news was made public, Lord Scarsdale said, with unconcealed pride, 'I begin to realise what a splendid position you have deservedly won. Congrats pour in from every quarter, and the country generally are as proud of you as I, your father, am, and more I cannot say.'[55] Curzon was overwhelmed.

In the glare of the public limelight from the time of the first announcement, Mary found, 'We begin to be treated like grandees'. If they travelled by train, 'station masters always meet us, carriage reserved, low bows and crowds staring.'[56] It was an exhilarating experience. Hitherto as an under-secretary Curzon enjoyed little position or publicity. Now suddenly a dazzling role had come within his grasp. As Mary exulted, 'From nobodies we have jumped into grandeur'.[57] Indeed they had.

Chapter 9

Fin de siècle *India*

On the night of Tuesday 22 June 1897 there were celebrations across India to mark the Diamond Jubilee of Queen Victoria. Halfway across the world, in London, setting out for the ceremony at Westminster Abbey, the 60 year-old Queen had sent out a message to the Empire: 'From my heart I thank my beloved people'. Westminster Abbey presented a magnificent spectacle, crowded as it was with splendid uniforms and beautiful dresses enhanced by a 'dim religious light'.[1]

In Poona, the summer capital of Bombay Presidency, the sumptuous Government House, Ganeshkhind, with its high gilded ceiling and gold and crystal chandeliers, formed the setting for Governor Sandhurst's banquet. The Governor had reported to the Viceroy, 'I have a feast of 100, and a reception on Jubilee night', later adding, 'Bonfires were lit on the eight highest points surrounding Poona and rockets were let off.'[2]

Night had fallen over the city in a thick blanket of sweltering haze as a line of carriages wound their way past the magnificent gardens to draw up at the door of the Governor's residence. Braving the heat, the guests came arrayed in all their finery—officers in full dress uniform emblazoned with decorations, the women in billowy dresses of silk and satin. When the guests had assembled at the appointed time the ADCs appeared escorting the Governor and his wife. Dresses rustled as the ladies curtsied. The orchestra struck up a tune and the party began.

Just a little distance away from this scene of revelry, parts of the native town of Poona looked as though they had been devastated by a cyclone. Entire tenements had been ransacked where desolate looking residents sat clustered in corners convulsed with fear. The deadly bubonic plague had broken out in the city, taking a heavy toll of the inhabitants. In no time the stricken were caught in the disease's vice-like grip. For bubonic plague, or Black Death as it was popularly called, was wildly infectious. In Bombay, where it had broken out earlier in the autumn of 1896, it had taken a toll of 9,000 lives in the first six months.

Lord Elgin's viceroyalty, which began in 1894 amidst what seemed like placid waters, found itself suddenly in the midst of a storm. From the onset the Earl of Elgin had been reluctant to shoulder the burden of India, believing himself not to be equal to the task. He was soon to have problems in asserting himself, both with the bureaucracy in India and the home government. The son of an earlier Viceroy who had died prematurely, the 9th Earl of Elgin had little of viceregal bearing. Photographs show him as a bearded, somewhat dishevelled but homely looking country squire who looks as though he would be more at ease lounging with his dogs and family than in the palaces he was called upon to inhabit. In appearance, bearing and action, Elgin could not have presented a greater contrast to his successor.

Alarm bells started ringing for Elgin when the south-west monsoon failed in the Allahabad division in 1895. By the new year the area was in the middle of a famine. The rains failed again the next year and that October brought 'the entire Indian continent face to face with the most widespread...and the greatest impending famine in the century'.[3] This 'grim spectre' covered 225,000 square miles and more than 62 million people in British India. The price of foodgrains escalated and grain riots broke out in Delhi and Bombay.

While taking steps to check the famine, Elgin had not how-ever allowed this 'grim spectre' to interfere with his tour of the native states. Interpreting this as callous indifference, Indian newspapers angrily cried out, 'there are on one side the wails of poor famine-stricken people, and on the other the sound of noisy festivities.' The Viceroy was bitterly criticized for indulging in 'revelry in stately edifices while hungry people cried for food...'[4]

Close on the heels of the famine had come the devastating plague. Lord George Hamilton, the Secretary of State for India in London, confessed that he was 'more concerned about plague than famine', explaining that a market once lost, or even partially diverted was not easily regained. For these reasons, Hamilton persuaded the Government of India to close the ports to Haj pilgrims.[5] Elgin had protested, saying that 'bitter experience' had shown 'that to enforce orders which the ignor-ant masses could regard as infringing religious privileges must be dangerous.'[6] But he was overruled.

Under pressure, the Government of India had been forced to abandon the municipal authorities and hand over plague eradi-cation to a special executive whose approach to the problem was nothing short of ruthless. The poor ignorant peasants of the Deccan having barely recovered from the ravages of the famine of 1896 were now asked to submit to plague regulations which frightened them almost more than the plague itself. Women were subjected to 'inspection', which was wantonly interpreted as an authorization for those concerned to infringe all taboos and to touch and handle the women in any way they chose.

The choice of Walter Charles Rand for the job could not have been more unfortunate. Reputed to be a tyrant and a bully, Rand had summoned British troops to his aid and had swept down upon the slums like the proverbial wolf to the fold. Plucking out men, women and children from their homes, he burnt their belongings, and desecrated their shrines. No explanations were offered, no answers given. Suspected victims were forcibly evacuated, their families coming to hear of them only after they were dead. Terror gripped the city. 'A small spark might cause an explosion', Elgin fearfully observed.[7] But once again, he was scarcely heeded. The Viceroy had barely been able to spare Bombay from the humiliation of 'corpse inspection', which the India Office had been so keen to enforce.

As resentment against Rand and his 'vast engine of oppression' smouldered over the Deccan, Bal Gangadhar Tilak, the Chitpavan brahmin Secretary of the Standing Committee of the Congress in the Deccan, angrily demanded, 'What people on earth, however docile, will continue to submit to this act of mad terror?'[8] Declaring that the official actions were trespassing on individual freedom, Tilak condemned Rand in a speech at a Shivaji festival on 13 June and published this in *Kesari* on 15 June, a week before Victoria's Diamond Jubilee. In the speech Tilak justified the killing of King Afzal Khan by the legendary Maratha chief Shivaji.

On the night of 22 June, as the guests were being driven home in their carriages from Government House, shots rang out into the night. In the round of firing, Rand and his assistant, Lieutenant Ayerst, were killed. The assassins' bullets missed Mrs Rand who was also in the same carriage. Before she could raise an alarm the assailants had fled into the night. Poona was promptly put under curfew and Tilak arrested on charges of sedition. He was sentenced to eighteen months imprisonment. The Natu brothers, suspected of complicity in the murders, were also arrested and detained without trial. Considered Sirdars of the Deccan, their ancestors had played a major role in enabling the British to establish ascendancy in western India.

Eventually the crime was pinned down to Damodar Chapekar, another Chitpavan brahmin, who said that he had been driven to the deed by the doctrines expounded by Tilak in *Kesari*.[9] In his autobiography, Damodar Chapekar confessed how he and his brother had waited with stolen pistols for Rand on the night of the murder. Poona was then swarming with people who had come to see the bonfires on the hill tops. Jostling with the spectators, the Chapekars had bided their time. At about 11 p.m. as the Governor's guests had begun to leave, they had jumped on top of Rand's carriage. Undoing the bottom of the flap, Damodar 'fired from the distance of about a span', the plan being 'to empty both pistols at Mr Rand so as to leave no room for doubt about his death'.[10] In their nervous excitement they also shot and killed the second occupant of the carriage—the unfortunate Lieutenant Charles Egerton Ayerst.

Surveying the scene on the assumption of his viceroyalty, Curzon, however, felt no reason to believe that the British were not fully in control and that India would not for always remain a British possession. He therefore loftily ordered the release of the Natu brothers, declaring, '...I fancy that the existence of a conspiracy itself, at any rate as a political movement, is now disbelieved.'[11] The new Viceroy went one step further, scolding the British authorities for not taking adequate heed of Indian pride and prejudice in combating the plague.[12] Curzon then ordered the release of Tilak before his prison term was complete. Valentine Chirol, Editor of *The Times of India*, complained of Tilak's release saying how it was '...considered in the Deccan as a fresh triumph. He was acclaimed by his followers as a "national" martyr and hero.'[13]

Until Curzon's arrival the aggressive politico-religious upheaval was confined largely to western India—Tilak's base of operations. In the early 1880s, the brahmins in the Deccan had been divided into two camps; one headed by the mild and cautious Justice Mahadev Govind Ranade who believed in opposing British policies through slow, constitutional means and the other by the fiery, young Tilak who did not frown upon incendiary tactics even though he owed allegiance to the Congress. In the battle royal that had raged betwen the two camps over control of the Poona Sarvajanik Sabha and the Education Society, Tilak had been eased out in the first round.

But Tilak's defeat was temporary. The introduction of the Age of Consent Bill in 1890, introduced to mitigate the evils of Hindu child-marriage by raising the minimum age of consent to twelve years, gave Tilak the opportunity to raise a storm of protest against foreign interference in Hindu religion. He appealed to the more orthodox adherents who had begun to view the growing westernization with unease. Alarmed by the violence of Tilak's propaganda, Ranade retired into the background and Gopal Krishna Gokhale[14] emerged as the new leader of the moderates.

Ten years younger than Tilak, the mild-mannered Gokhale could not have presented a more marked contrast. Slim, slight, with eye-glasses and a somewhat hesitant air, he gave the impression of a rather myopic and harmless professor. A believer in the ultimate fairness of India's rulers, Gokhale shunned violence and coercion. In 1905 he was to be elected president of the Congress. Tilak, on the other hand, with the help of the orthodox Natu brothers, had carried his violent propaganda into schools and colleges against the wishes of the moderate members of the Congress party. Encouraging the growth of gymnastic societies where training in the martial arts was imparted, in 1896 Tilak organized the Shivaji festival; the festivities, which extended over a week, providing a forum for discussions on Indian culture, nationalism and religion. Two years earlier he had revived the Ganapati festival, resurrecting the religious deity in a bid to arouse national pride and Hindu unity.

Curzon was not unduly alarmed. Much of the religious emotionalism and incendiary speeches were confined to the Deccan. Until the British arrived the Marathas alone had challenged the Moghuls. It was not unnatural that they transfer their hostility to the British. Not unaware of the problems before him, Curzon believed they had a way of dwindling. The Government of India generally had the support of the nationalist leaders, whose attitudes were moderate. The fourteen year-old Indian National Congress at the time of Curzon's arrival had little money, less sustained activity and no permanent organization. The fact that the early Congress leaders should have so largely entrusted their political affairs to an Englishman, could not have enhanced their credibility in the eyes of the government.

Alan Octavius Hume directly dominated the Congress from its beginning in 1885 to 1892, when he retired as General

Secretary to return to England. Hume, an ICS official, had served in the North-Western Provinces and was held responsible for the killing of more than 100 rebels during the 1857 Mutiny.[15] He had risen to be a Secretary to the Government of India in Viceroy Mayo's time, only to be superseded eight years later with Lord Ripon's coming to power. Hume worked his way back into the Government's good books, becoming the Viceroy's adviser on Indian opinion.

The Congress sessions of 1898 in Madras, 1899 in Lucknow and 1900 in Lahore had difficulty in finding funds for financing its conventions. After the death of the Maharaja of Darbhanga in 1898, few landlords came forward with financial help. Western India was able to attract some funds, especially from the rulers of Baroda, Junagadh, Bhavnagar and Gondal. The Congress was only galvanized into action immediately before and after its annual meetings and did not even have a constitution before 1899. Its reins were tightly held by the moderates—W. C. Bonnerjee, Pherozeshah Mehta, Surendranath Banerjea, Dinshaw Wacha, Sir W. Wedderburn and Gopal Krishna Gokhale— who Curzon believed were not capable of setting the Ganga on fire.[16] Romesh Chandra Dutt as President had welcomed the Viceroy on his arrival[17] and Curzon told the Secretary of State how he was '...sensible of the desirability of conciliating public opinion in India.'[18]

But outside the moderate leadership still waters ran deep. An aggressive new nationalism based on the militant affirmation of Hindu culture was sweeping the land, elbowing out the moderates who had carried forward the theories of a blameless English political life, British credentials having received a severe setback with Lytton's disastrous viceroyalty followed by the ugly controversy over Ripon's Ilbert Bill. Determined to establish British authority in Afghanistan, the unfortunate Lord Lytton had arm-twisted the Amir into accepting the presence of a British Resident at his court. Within months, the hapless Louis Cavagnari had been murdered in Kabul, forcing the government to mount another expensive and fruitless campaign.

In the Midlothian campaign that swept the Liberals to power in England, Gladstone had vowed that he would try to wipe out the sins of the Disraeli–Lytton Government and his choice of Ripon seemed to augur well for India. Though beginning with the repeal of Lytton's Vernacular Press Act, Ripon, however, soon ran into rough weather when, heeding the request of Indian ICS Behari Lal Gupta, he sought to introduce the Ilbert Bill. Infuriated at the idea of Indian judges trying European subjects, the white community had literally caught the Viceroy by the throat and threatened to deport him back to London! As the mild Ripon crumpled under the attack, educated India, shaken to the core noted,

> It burnt in the minds of the Indian politician the fateful lesson that if India is to protect her liberties and secure an expansion of her legitimate rights, she must initiate as violent an agitation as enabled the European residents in the country to compel the Government of Lord Ripon to practically throw out that proposed measure.[19]

In the wave of disillusionment sweeping the country the new literary renaissance quickening in Bengal found willing listeners. In Calcutta, poet, novelist, dramatist Bankim Chandra Chatterjee aroused a pride of race in young men by exhorting them to lay down their lives in the service of the motherland as symbolized by the goddess Maha Kali. His novel *Anand Math*, published in 1882, which told the tale of young men and women leaving their homes to fight for their enslaved mother, churned patriotic sentiments. His opening song 'Bande Mataram' rapidly became the anthem of Hindu nationalism. An early Hindu revivalist, Bankim Chandra gave a romantic tinge to militant nationalism, but his blood-curdling pleas did not shun violence. Young men thrilled to his passionate appeals and on reading his novel *Durgeshnandini* one said, 'Our sympathies were entirely with Birendra Singha, the Hindu chief of Gad-Mandaran; and the court scene, wherein the Moslem invader was stabbed through his heart by Bimala, one of the wives of the chief of Gad-Mandaran, made a profound impression upon my youthful imagination.'[20]

Soul-stirring musicals and theatrical performances glorified the lives of Indian heroes. The Bengali stage was gathering great momentum. When the play *Neel Darpan*, depicting the exploitation of poor peasants by British indigo planters had been shown on the stage, '...the audience got wild with passion against the White Planters; and sometimes they so far forgot themselves that they threw their shoes at the poor actor on the stage.'[21] During the Prince of Wales visit to Calcutta in 1876 when a High Court pleader, in his enthusiasm to ingratiate himself with the royal guest, arranged a 'purdah party', an aghast Calcutta society was up in arms. Upper-class Hindu women then rarely appeared in male company unless they were related by blood or marriage. A scathing satire *Purdah Party* was promptly mounted on stage to publicly humiliate the unfortunate lawyer. The Government moved to repress its performance.

Patriotic songs openly spoke of foreign exploitation. Soul-stirring musicals and theatrical performances further fanned the resentment:

> Sing, O my clarionet! Sing these words:
> Everyone is free in this wide world,
> Everyone is awake in the glory of science,
> India alone lieth asleep!
> China and Burma and barbarous Japan,
> Even they are independent, they are superior,
> India alone knoweth no waking![22]

A tide of reaction was setting in towards the social reform movements of the earlier part of the nineteenth century, and strenuous efforts were under way to wean the anglicized Bengali youth back to the fold. Rajnarayan Bose and Nabagopal Mitra had started the Hindu Mela in Bengal with the sole aim of counteracting the alienating effects of English education. At a time when the populace was obsessed with perfecting their hold over the English language, Nabagopal turned to the vernacular, proudly saying, 'English is not my mother tongue, and though I may use the vehicle of this foreign language...I feel no call to waste my time and energy, in trying to master the senseless idiom of it.'[23] His gymnastic schools became training academies in martial arts teaching Bengali young men to wield the *lathi*, the dagger and the sword. The wealthy zamindar, Debendranath Tagore, father of poet Rabindranath, lent his support to the Hindu Mela.

In 1893 the diffident, and not-so-successful Gujarati barrister Mohandas Karamchand Gandhi had set sail for South Africa to represent a Muslim firm. When travelling in a first class train compartment in Natal, a white man had peremptorily ordered him to leave. The rest of the story is well known. Born in 1869, Gandhi was ten years younger than Curzon. *Fin de siècle* India saw eleven year-old Jawaharlal, only son of wealthy lawyer and Congressman Motilal Nehru, being brought up in the luxurious, anglicized ivory tower of Anand Bhavan in Allahabad where his Irish tutor, F. T. Brooks had begun introducing him to the theosophy of Mrs Annie Besant.

Hindu organizations had overnight begun to attract young men and to galvanize them into action. When the fiery young Swami Vivekanand returned after his glorious success at the Parliament of Religions in Chicago in 1893 he was given a hero's welcome. Vivekanand told his followers, 'The one common ground we have is our sacred traditions, our religion...In Europe political ideas form the national unity. In Asia religious ideas form the national unity.'[24] By 1896 Vivekanand had founded the Ramakrishna Mission. The Suddhi, or ceremonial purification movement of the Arya Samaj, was also giving Hindu revival its militant thrust, providing a powerful weapon for the new generation of Hindus humiliated by their inability to counter the attacks of Muslim and Christian propagandists against their religion. Though the Arya Samaj had failed to take root in Bombay, the message of the Gujarati brahmin was finding a warm response in the Punjab. B. C. Pal working in Lahore reported:

> In the message of Pandit Dayananda they [the Punjabi Hindus] discovered first, a powerful defensive weapon by which they could repudiate the claims to superiority of Christianity and Islam over their national religion. Dayananda, in the second place,...made a violent attack on Christian and Moslem propaganda showing up the unreason of both these alien systems![25]

Militant Hinduism's protection of cows and the consequent emergence of Gorakshini Sabhas threatened communal trouble and riots broke out in the North-Western Provinces, Bihar and Bombay in 1893, provoking Viceroy Lansdowne to predict, 'The subterranean connection which has now been established between the Congress and the Cow will, unless I am mistaken, convert the former from a foolish debating society into a real political power, backed by most dangerous element in the native society.'[26]

Secret societies vowing to destroy British rule flourished. In Calcutta, poet Rabindranath and his brother Jyotindranath belonged to them. Inspired by Surendranath Banerjea's stirring lectures on Mazzini and the Italian freedom movement, one youthful member said, 'We saw or imagined a great similarity between the position of the Italians under Austrian domination and our own position under British rule'. The Italian Carbonari had advocated assassination in the attempt to win freedom. These societies had their own elaborate initiation vows; in one, members signed their pledge with blood drawn from the breast by the tip of a sword. But, as yet, the youthful agitators in Bengal were in their own words, 'dreamers of wild dreams... whose thought and imagination alone were of a revolutionary character, but who never seriously meant to rise in physical revolt against the British authority in the country, or who hoped to secure the emancipation of their people by a campaign of political assassinations.'[27]

Lean and intense, with mesmerizing eyes, twenty-one year-old Aurobindo came to the same bitter conclusion when he arrived in Calcutta in 1893. Newly returned from Cambridge, Aurobindo seethed with a marked hostility to the British. Having been sent by his father to study in England at the age of seven, Aurobindo had quickly mastered Greek and Latin and won a scholarship to King's College, Cambridge. At the age of 18, he had effortlessly passed the entrance to the ICS but was eventually disqualified for failing to appear in a riding test. He had lived in England so long that he found it difficult to express himself in Bengali but when he returned to India he began to plan secret societies for armed insurrection against the British. In England he had been deeply influenced by the French revolutionary movement and its 'purification by blood and fire'. In London he had helped to form a secret society called the *Lotus and Dagger*. But in Bengal, to his dismay, he found only apathy and fatalism. There were no serious plans for evicting the British. Disgusted, he left for Baroda to work for the Gaekwar, grumbling, 'The Congress in Bengal is dying of consumption... its leaders, the Bonnerjis and Banerjies and Lalmohan Ghoses, have climbed into the rarefied atmosphere of the Legislative Council and lost all hold on the imagination of the young men.'[28] As the nineteenth century drew to its close, young Aurobindo felt marooned, depressed and defeated.

Moderate nationalists still had faith in the providential British connection and were said to feel 'no sense of anti-climax in the combination of passionate lectures of Mazzini and Garibaldi with pleas for more jobs in the ICS or more seats in the Councils.'[29] Dadabhai Naoroji, to be thrice President of the Congress, was loyally saying, 'Nothing is more clear to the heart of England—and I speak from actual knowledge—than Indian welfare; and if we only speak out loud enough, to reach that busy heart we shall not speak in vain.'[30]

The Indian Councils Act of 1892, which enlarged the size and function of the legislative councils had been welcomed in India even though it failed to satisfy the Congress demands for the election of at least half the members. Curzon was not unaware of these demands, as Under-Secretary of State for India, he had piloted the Bill. In the first election under the new Act, out of the six people elected to the Bengal Council, five were Congressmen. The Congress demand for simultaneous Civil Service examinations had yet to be met.

Sir John Gorst, Under-Secretary of State for India told Lansdowne, 'Nothing could be more mischievous than the crude application of British democratic maxims to India...But between Scylla and Charybdis there is a safe passage.' And he advised Lansdowne to avoid '... on one side stupid resistance to all change, and on the other weak surrender to fantastic theories.'[31] As though taking the cue, Curzon had begun his rule by upbraiding his own people over their clumsy handling of the anti-plague measures. 'I fancy that almost everywhere we began wrong,' he admitted with rare humility and under-standing. 'In those days it was all science, and compulsion and evacuation at the point of a bayonet ... You cannot do this sort of thing with large aggregations of Asiatics.'[32]

Chapter 10

'We Might as Well be Monarchs'

In December 1898 the Curzons sailed for India on the S. S. *Arabia*. Earlier Curzon had been fêted at a Court Ball and ceremonials at Oxford and Southport. In October, 200 Etonians had met at the Cafe Monico in Piccadilly to bid Curzon farewell, with former Prime Minister Rosebery declaring in his toast that a great deal more than the Battle of Waterloo had been won on the playing fields of Eton.[1] In tones hushed with reverence Curzon had said, 'The East is a university in which the scholar never takes his degree. It is a temple in which the suppliant adores but never catches sight of the object of his devotion. It is a journey the goal of which is always in sight but is never attained. There were always leaders, always worshippers, always pilgrims. I rejoice to be allowed to take my place in the happy band of students and wayfarers who have trodden that path for a hundred years.'[2]

To the outgoing Viceroy he had confided, almost prophetically, that there was some incongruity in all the wining and the dining. No doubt 'it is all very generous and very encouraging. But surely the entertainment and congratulations ought to come after, not before, the performance?'[3] Travelling with the Curzons were his private secretary and friend from Balliol, Sir Walter Lawrence, the Military Secretary, Colonel Sandbach, and two ADCs—Lord Suffolk and Captain Meade.[4] On Christmas Day the ship halted in Aden and Mary joyously noted this was 'the beginning of our kingdom'. They were taken ashore on a white state boat 'rowed by eight blacks in red liveries'.[5] Bubbling with excitement she told her father:

> On Friday we land at Bombay . . . I will send you papers of all the doings . . . George & I drive in a state carriage with four horses, postillions outriders & escort, and behind two syces running holding parasols over our heads . . . all this time salutes of thirty-one guns will be firing & all the warships in the harbour cannonading—so it will be thrilling . . . Harmsworth of the *Daily Mail* has two correspondents in this ship . . . So we have the search light of publicity full upon us![6]

In Bombay the Curzons got their first taste of the pageantry that was to accompany them throughout their stay in India. The landing was covered in crimson carpet. At the end of the landing the state carriage was drawn up and Mary reports, 'George and I got in to "God save the Queen", and drove through seven and a half miles of packed streets . . . bowing & smiling hard as we ever could!'[7]

It was a novel experience and exhilarating. The pomp and protocol took Mary's breath away and she made no secret of this. That evening the Governor, Lord Sandhurst, gave a reception for 1,400 people and as the loyal subjects of the Crown—'English—Parsee—Mohammedan & Hindu filed past', the Curzons standing beside their gold chairs on a gold carpet thrilled with pride.[8] The next evening at 5.30, they drove in state to the station accompanied by their escort. Mary who found it difficult not to talk in superlatives said, 'The station is the finest one I ever saw in my life'. Troops were drawn up and red carpets lined the floor. They walked to their special train, a band playing 'God save the Queen', the troops presenting arms. Mary added gleefully, 'The general traffic manager was on the train to look after the Royal family!'[9] It could not but seem to them that '. . . for all the fuss & ceremony we might as well be monarchs'.[10]

At exactly 4.50 in the afternoon of 3 January, the state carriage bearing the incoming Viceroy and his consort drew up at the foot of the grand staircase of Government House in Calcutta. The stately procession had driven past Dalhousie Square between lines of troops; the viceregal carriage accompanied by the fabled bodyguard—tall, handsome, hand-picked troopers resplendent in their livery of scarlet and gold. Mary had never seen such crowds as there were that cold January afternoon lining the streets of Calcutta, straining to see the new Viceroy. Every balcony, every window above the street was jam-packed with people. The Curzons had 'bowed & bowed & all cheered and it was a *marvellous* sight.'[11]

At Government House the enormously wide flight of stairs, used only for such ceremonials, dazzled with the resplendent uniforms of naval and military officers, judges, civil servants, dignitaries of the church and the Indian princes. In accordance with strict protocol, at the foot of the staircase stood the Lieutenant-Governor of Bengal, at the top the outgoing Viceroy to hand over to his successor the brightest jewel in the Crown.[12]

The glittering array of guests included the great Maharajas. Patiala, Scindia of Gwalior, Kashmir—'in *magnificent* jewels' attracted the special attention of the new vicereine-to-be. 'No. 1 had a breast-plate of *enormous* diamonds & necklace after necklace,' she gasped. 'No. 2 had emeralds to make your eyes water & No. 3 had twelve rows of pearls as fine as Consuelo's *biggest*!' Wistfully she added, 'My fine feathers paled before this magnificence.'[13] Mary was presented with the humbling but reassuring thought that it was her husband who was to control the destinies of these magnificently bejewelled men.

For Curzon, as he set foot on the great red-carpeted staircase of Government House, it must indeed have seemed a journey to heaven: not merely a moment of great triumph but also the fulfilment of a cherished dream. When he had first entered the portals of Government House twelve years ago his thoughts had been spurred to returning one day as ruler. Today he had achieved his ambition and was the centre of a historic event. As he made his slow, stately, solemn upward climb, the cannons from the nearby Fort William were set off in a 31-gun salute. The dramatic ascent from a junior minister to Victoria's sole representative on the Indian subcontinent was to have far-reaching consequences for Curzon, shaking an already shaken equilibrium.

Painfully young, somewhat artless, a little vain, Mary was not to help matters. Her upbringing in the Leiter household had hardly prepared her for the great position she was now going to hold. Later, St. John Brodrick was to harshly hold her and her American background responsible for much of Curzon's problems in India. 'Lady Curzon, though possessing brilliant qualities, had not been brought up in England and had none of the traditional knowledge that many English women possessed of the "give and take" of public life,' he said.[14]

For three days after their arrival, Lord Elgin remained Viceroy and Curzon his guest. In appearance they could not have posed a greater contrast—the short, bearded, lackluster Elgin was dwarfed by his imposing, statuesque successor. A monograph of Elgin says, 'though not yet forty-five when he came to India, he looked considerably older; he was careless about his dress and avoided wearing uniform if he could help it.'[15] Lady Elgin was marginally better, being '...possessed of more glamour than her husband; though she could not adequately fill the place of Maud Lansdowne [her predecessor].'[16] In appearance Mary Curzon far outshone them both.

On the third morning after his arrival, Curzon was proclaimed Viceroy in the Council Chamber in the north wing of Government House, just prior to the Elgins's departure. 'They [the Elgins] soon came down "en masse" for goodbye,' said Mary, somewhat patronizingly describing, 'Lady Elgin, two unmarried fat girls...Lord Elgin and their Staff.' Curzon accompanied them by carriage to the landing stage at Princep's Ghat where they were to sail off on an Indian ship. Going across to the now empty Council Chamber, Mary had watched them drive off, not failing to notice that 'Lord Elgin [was] still taking the right side of the carriage and taking the salute' whereas, her husband, now the Viceroy, sat 'by his side modestly doing nothing'. Yet, in Mary's eyes, it was Curzon who was 'towering three feet above the small form of Lord Elgin.'[17]

At the turn of the century a Viceroy's position was awesome. Even though he had to report to the Secretary of State and the India Council in London, Calcutta was 7,000 miles away and the journey took seventeen days. Removed from Westminster, Curzon began to feel that he was indeed the monarch of all he surveyed.

By 1899 Calcutta had developed into a home away from home which was forever England. In the 200 years since Job Charnock had stopped on the eastern bank of the Hooghly, armed with a licence to trade from the Emperor Aurangzeb, the British had built for themselves a metropolis which was European in architecture, style and morals. As in London, Calcutta had its vast open park—the Maidan—at its heart; the Hooghly river breeze brought along with the salt air, as did the Thames, the aromas of spice and tea.

The British built St. Anne's Church in 1709, the Fort in 1712 and the first theatre opened in Lalbazar in 1745, by which time the European town was about a mile long and two furlongs wide, with the Indian town squeezed to the periphery beyond. The Old Court House (or the old Town Hall) came up in 1727, its subsequent enlargement having been financed by an Indian, Omichand, who contributed Rs 20,000.[18] Daniel's drawing shows elephants sauntering in the nearby Tank Square area.[19] The Asiatic Society was founded by Chief Justice Sir William Jones in 1784. He also began the building of the gracious suburb of Alipore near Garden Reach in the south of Calcutta. St. John's Church opened with much fanfare on Easter Sunday in 1789. A contemporary account said, 'A Hindu [Maharaja] Nabakissen' contributed towards the building of the Church by donating...a piece of ground, valued at 30,000 rupees.'[20] By 1803 Government House had been completed and by 1813 so had the new Town Hall. The Strand Row on the embankment, built from funds collected through a Lottery Committee, was finished in 1824. Parallel to it ran Clive Street, 'the grand theatre of business' with practically 'every public mart in it'.[21]

By the turn of the 19th century Calcutta was a city of palaces. Huge white neoclassical mansions stood amid carefully cultivated gardens facing wide boulevards. The smaller buildings were also painted and ornamented to resemble their larger counterparts. Public buildings had been created to appear awesome, the grandiloquent style heightened by the Palladian façades and columned doorways. Massive porticos accommodated no less than two carriages abreast. But behind these immense villas were the invariable huddle of dingy mud huts, for the European could scarcely do without his battery of domestic servants.

The old families of Calcutta lived in equally imposing mansions to the north of the European settlements; most of them having acquired positions of power and influence through collaboration with the British. The Tagores, the Debs, the Basaks, the Dattas, the Setts traced their origins to Gobindpur which along with the villages of Kalikata and Sutanati had merged to form Calcutta. The British dismantled Gobindpur by an imperial fiat to make a way for the building of the new fort in 1757.[22] The Setts, the merchant-weavers who had supplied cotton goods to the Company, moved their family deity, Gobindji, to a spot in Burra Bazar in the north. The others followed. Sobharam Basak, a mid-18th century merchant, had 37 houses in Burra Bazar besides ponds and gardens in north Calcutta. Dabbling in a variety of money-making activities from trading in cotton, spice and opium, Sobharam found time to build the Jagannath temple. Devoted to religion, Sobharam was said to have left Rs 37,000 to be given away in charity upon his death of which Rs 3,000 was earmarked for school teachers.[23]

The wealth and style of the Indian merchant princes was legendary. Nemaicharan Mallik enjoyed a fortune of around Rs 3 crores at the turn of the 18th century. Upon his death he left 96 premises in Calcutta and Rs 24 lakhs in cash. No computation was possible of the value of his jewels. His woollen shawls were said to be so precious that his eight sons tore them into precisely eight pieces each! At the *shradh* (or post-funeral feast) in the Nemaicharan Mallik household, villages within a radius of thirty miles emptied out as their inhabitants made their way to Calcutta to partake of the Mallik largess.[24]

At the time of Curzon's arrival the total wealth of the Tagores was said to be in the region of Rs 1,000,000.[25] Dwarkanath Tagore, grandfather of the poet, starting life as the dewan of the East India Company's salt and opium department, rose to become proprietor of Carr, Tagore and Company, to sit on the board of the Union Bank alongside European directors and to live and entertain in royal style at his fabled palace at Jorasanko. In 1838, Sir Dwarkanath founded the loyalist Landowners' Association. His son, Debendranath, financed his factories which manufactured silk and indigo from the rent of his father's estates but shunned the luxurious life-style. When the British Indian Association was started in 1851 to defend the rights of Her Majesty's Indian subjects, Debendranath became its first secretary and the Sanskrit Scholar, Radha Kant Dey, its first president.

Radha Kant's ancestor, Raja Nabakrishna Deb had amassed immense wealth as a Political Banyan to the East India Company, starting his career as a teacher of Persian for Warren Hastings. In 1774 he was given large tracts of land in Sultanati. Here he built the palace where the Deys fêted Swami Vivekanand upon his return from America. Each Durga Puja brought hundreds of visitors. One invitation from the old Raja boasted that 'Sahib [Robert Clive] will be present'.[26]

Hem Chandra Mallik, scion of the Basu Mallik family whose fortune had been made from a partnership in the ship-building firm of Beauchamps and Company and later from the Hoogly Dockyard, presided over the palatial family mansion at 12 Wellington Square. The U-shaped house with a garden in the middle and a park in front had been designed in the European style, boasting a gleaming wooden dance floor and a billiard room. The reception rooms were strewn with carpets, the ceilings festooned with chandeliers and the walls adorned with paintings—the most treasured being a painting of George Washington by Gilbert Stuart. The elegantly furnished bedrooms were wallpapered in pink and mauve. European teachers came to teach the daughters English and to play the piano. Downstairs, in a separate kitchen, western delicacies were prepared for Europeans and anglicized Indian friends. A menu for dinner on 24 November 1897 reads:

Soupér
Mayonnaise de Saumon
Poulets piqué Rôti et Langues
Saucisses Rôlês
Paté de foie gras en Bellevue
Dinde farcies aux Truffées
Jambon d' York
Sandwiches Varieé
Bignolias à la Crême
Meringues à la Chantilly
Gelée à la Macedoine
Gateaux Fantasie
Patisserie et Confisserie Assortis
Glace, pezzi duri
Cafe Noir[27]

Westernized, English-speaking, urbane scions of Bengali zamindari families were members of the British Indian Association. The two communities co-operated at work and in business but, as the 19th century drew to its close, the occasions for the twain to meet socially were becoming more and more infrequent.

The Bengali élite were, however, included in the invitations sent out from Government House. Mary reports one occasion when 'there were Natives galore who surrounded George so that he couldn't meet nearly as many English as I did'.[28] The Curzons hob-nobbed with Indian princes, staying with them when the Viceroy toured the Indian states. But when in Calcutta they did not deem it necessary to fraternize with the Bengali.

Was this due to there being other more pressing engagements, or was it that Curzon was not able to overcome his initial prejudice towards the Bengali on his first visit to Calcutta, in 1887, when he had called them an uninspiring and unmanly race.

His predecessor, the languid Lord Dufferin, was known to happily respond to Bengali hospitality. The charming and cultivated Lady Dufferin described with great relish her participation in 'a native dinner party' where, wholly throwing herself into the entertainment, she changed into a saree and sat on the floor to enjoy the Bengali meal.

> As soon as we got to the house, we were divested of our own gowns, and were draped in sarees...The floor table was very large...One large silver dish, with a pile of rice in the centre of it, was put before each person...and I was afterwards much complimented on the ease with which I managed my dinner, seated like a Buddha on a mat. I was told that I did it 'as well as any Bengali,' ...food had to be manipulated with unaided fingers...we all enjoyed this dinner very much.[29]

Mary who has kept copious notes of her encounters in India does not, however, mention a single visit to a Bengali family. In fact, the Curzons, conscious of their position, paid no informal visits. Mary moans to her mother, 'George and I are so bottled up that we can never go to any private house to an entertainment'.[30]

European Calcutta did its conscientious bit towards Indian charities. Over Christmas of 1898 Calcutta hosted a Bazaar in aid of St. Mary's Home at the Town Hall complete with twelve stalls of handicrafts, a band, and an inauguration by Lady Woodburne, wife of the Lieutenant-Governor of Bengal.[31] The new Vicereine was also being called upon to do her own bit toward ameliorating the lot of the *ryots*. The end of January found her visiting two hospitals and a nurses' hostel. 'Colonel Fenn [a doctor on the Viceregal staff] and Lord Suffolk [ADC] went with me, and Colonel Harris, the Head of the Medical College, took us all over that Hospital, and Dr Joubert took us over the Eden Hospital, and Miss Taylor took us over the hostel for nurses who are training for the Dufferin Fund,' she reported.[32] Much more pleasurable, however, for the young and beautiful Vicereine was the Fancy Dress Ball in aid of the Women's Friendly Society. Wearing her Empire gown, a flushed and elated Mary recalled how on her arrival at the venue '...all the Swells were standing in two lines through which I walked bowing'.[33]

European Calcutta lived behind its cloistered walls, though within this charmed circle, there were precise dividing lines. 'Members of the Civil Service were very exclusive, holding themselves much more aloof...the military formed another distinct set; while the mercantile people, lawyers, barristers, and others not in any government service, had their own particular circle,' recorded a British resident of Calcutta at the turn of the 19th century.[34]

European Calcutta went on through its unchanging daily pace. Morning began with an early morning ride on the *maidan*; for the more privileged there would be the occasional hunt and Mary was soon reporting, 'George went shooting at 7 a.m.'[35] Breakfast or *chota hazari* was an elaborate meal of fish, mutton chops, cutlets or other dishes of meat, curry and rice, bread and jam and lots of fruit—oranges, plantains, lichis, pineapples, papayas or pumelos—according to season. Some drank tea, but most had iced claret and water.[36]

As the men went off to work, the women absorbed themselves with visiting friends or shopping. The young Vicereine was denied these feminine diversions though sometimes she went to Hamiltons to see the 'jewels'.[37] Hamiltons of Old Court House Street, a stone's throw from Government House, enjoyed the patronage of viceroys, wealthy British nabobs and princely houses of India. No grand wedding or party in Calcutta was complete without a shower of Hamilton brooches, necklaces and pendants. By noon the ladies returned home, lunch or tiffin being at two in the afternoon. After a meal of soup, hot meat, curry rice, cheese and dessert, washed down by claret and beer, there was siesta under the *punkah*, the old hand-pulled variety giving way to the electric fans. At five o'clock everybody went off to 'take the evening air', the most popular spot being the Strand, famed since the days of Warren Hastings as a favourite promenade by the riverside. Even the viceregal couple rode by occasionally. Mary reported, 'At 5 he [Curzon] came back for me and we went out to drive'.[38] No doubt they ran the risk of running into 'native India' and Mary was once taken aback by the sight of '. . . a fat Native . . . driving in a landau with both windows closed: his brown skin glistening through the panes of glass'.[39]

As the year 1898 drew to a close the European society of Calcutta was busy going through its staggering round of 'cold weather' engagements. There were dinner parties and garden parties, subscription balls, billiard-playing, polo, theatre, opera, horse-racing, cricket and steeple-chases. In the evenings the music-lovers trooped to hear the Calcutta Orchestra Union's rendering of *Lucrezia Borgia* at the Town Hall. The Opera House on Lindsay Street, on the opening night of *Little Minister* had experienced a crush: the Viceroy and the Countess of Elgin had been the chief guests. With its unpretentious wooden façade, the Opera House did not present an imposing spectacle, but the crowd it drew was select. In 1876 the visiting Prince of Wales had specially gone there to see the comedy *My Awful Dad*. Theatre Royal on Chowringhee also drew big crowds for its show *Hudson's Surprise Party*, Miss Ivy Scott bringing the house down with her song 'I do love you'.[40] The 'heaven born' had their customary annual dinner on 30 December. One hundred and three civilians assembled in the banqueting hall of the Town Hall; Mr H. E. Cotton, Chief Commissioner of Assam gave the toast of the evening.[41]

Traditionally, the finest entertainment was given at Government House. On 26 January Curzon wrote home, 'It is three weeks since I assumed office tomorrow. Within that time we have had Levee, Drawing-Room, State Ball, State Evening Party, three big dinners of sixty and four more of 20–30, so we have not been idle.'[42]

Being the seat of power, Government House dominated Calcutta society by its life and gossip. The mansions of the civilians, the army generals, the maharajas and the merchant princes throbbed to the rhythm that emerged from the viceregal household. A week of excitement could follow a nod, a smile, or a word sent by the Viceroy in a certain direction. Invitations to functions at Government House were more precious than gold.

What a delightful change all this was for the Curzons after their dull, monotonous, years in London. As Under-Secretary, Curzon had not had many invitations come his way. Mary had complained to her parents how 'London life is a continuous striving, striving, striving to keep going, the little people praying to be noticed by the great, and the great seldom lowering their eyelids to look at the small'.[43]

The Viceregal palace in Calcutta had cost the East India Company £63,291, a sum which the directors thought shockingly extravagant and a sufficient justification for recalling the Governor-General. Wellesley's desire had been to build a residence worthy of the highest representative of British supremacy in India, to give an outward appearance of the power and pomp of the Empire. Government House was indeed imperial. Tall windows gazed down on terraces, statues, gardens and ornate carriages drawn by magnificent horses. Inside, the gleaming halls and tall shaded rooms were gilded with marble, mahogany, gold, crystal, velvet, silk and, occasionally, simple homely chintzes. Beneath huge chandeliers, rich oriental rugs lay on marble and parquet floors.

To guard this paradise, to look after the guests, tend its lawns, pick its flowers, groom its horses, polish its carriages, clean its floors, make its beds, repair its curtains, serve its banquets, bathe and feed the viceregal children, took several hundred human hands. The Viceroy's bodyguards alone ranged anywhere between 120 to 400 men. Lord Elgin had reported to Curzon about the vast establishment that awaited him there. 'It is a very big concern for a census just taken in Simla showed as establishment to number 700 in all.'[44] This army of soft-footed salaaming servants, fussing over their masters with many attentions, heightened the illusion of royalty.

A most 'inconvenient house', Mary complained deprecatingly. Her bedroom was so huge that half of it had to be curtained off to make a sitting room. But she could not but have relished the luxury of servants. When she wanted a bath, one man heated the water, another got the tub, a third filled it, a fourth emptied it.[45] What a relief after her housekeeping efforts in London

The 'bird cage' lift, Government House, Calcutta. Curzon installed the lift as part of his modernization programme. Drawing by Desmond Doig. *Courtesy The Statesman Ltd.*

where she had felt 'English servants are *fiends*'. She had then wept that they were 'malignant and stupid and make life barely worth living'.[46] In England, when her father-in-law Lord Scarsdale had come to stay, things had been brought into a state of confusion. 'It was the signal for everything to go wrong. His fire is forgotten and goes out; no coal in his scuttle; all the meals late, a roast of mutton proves bad; the breakfast so nasty nobody could eat it'.[47] Now a veritable host of *khidmatgars* in scarlet and gold waited at her dining table, one behind each chair, whether it was a dinner for ten or 120 people; senior servants were distinguished by gold embroidered breast-plates, the lower orders sporting monograms of the Viceroy embroidered upon their chests.

Over weekends, the Curzons escaped from Calcutta to the Viceroy's country house at Barrackpore, fifteen miles up river. In a 350-acre green and shady park in Barrackpore, a succession of imperial rulers and their wives had created an isolated, miniature world, as fantastic as a precisely ordered mechanical toy. Outside this imperial park, the Viceroy's bodyguard rode day and night on ceaseless patrol. Inside were monuments—a Gothic church, a menagerie, domed lodges, bungalows for the sentries with pedimented doorways, an artificial Gothic ruin, Italian gardens with balustrades and a bandstand, and studded with velvet lawns. An artificial lake at the north front of the house had an Arcadian bridge. At one end of the lake stood a Corinthian temple built in memory of war heroes in the Java campaign; not far off was Lady Canning's tomb lit by an eternal flame.

At Barrackpore, the viceregal style was no less elaborate . A vast army of servants and staff moved with the Viceroy from Calcutta, even when the Curzons went for a mere weekend. Life for the Curzons was easy and informal at this suburban retreat. Society had long since abandoned Barrackpore. Since Simla became the official summer capital, few Europeans thought it worth their while to maintain separate establishments at Barrackpore as well.

Free from the official formality of Calcutta, Mary spent her days peacefully sitting under a giant banyan tree 'covered with creepers and orchids that hang in festoons from its pillared arcades'. It provided a perfect summer house. Here Mary sat and wrote letters watching the native river craft lazily winding their way up the Hooghly. Their two baby daughters, Irene and Cynthia, were wheeled round the park in their rickshaws happily clutching their dolls.[48] Lunch was served outdoors under the giant banyan tree, with the liveried servants hurrying to and fro from the house. Somewhat later in the evening, after dinner, Curzon took time off to play billiards with members of his staff while Mary read by lamplight. Everyone except the Viceroy went to bed early, the river breeze lulling them to sleep.

A quarter century later, Curzon remembered with great nostalgia:

The journeys up and down the river in the twilight of Saturday evening, or in the dewy radiance of a Monday morning. Motors had not yet invaded Calcutta, and a small steam launch was the only means of transport. To leave the city in the late afternoon and after tea on the deck, to live in a lounge chair and watch the changing panorama of the river banks as they flew by—the thick fringe of vegetation and the feathery palm-tops; the smoke of the native villages; the white-clad figures moving up and down the dilapidated ghats; the glare of the electric light suddenly switched on in some great jute mill, from which the throb of the engines hurtled across the water; the vast bulk of some huge flat, laden with tea or jute from the interior, sliding noiselessly down the full current; the peeping white fronts of villas, once the garden houses of Europeans, but now deserted by them; the rows of crumbling Hindu shrines on the river's edge, or the fantastic towers of some Pagoda silhoutted against the sky; on the one side the gathering dusk, on the other the red sun sinking in blood on his funeral pyre; and then, when light had vanished and all was swathed in shadow, to land by the glimmering tomb of Lady Canning, and to walk up the gravelled terrace to Barrackpore House, the hand-borne lanterns twinkling in the darkness ahead—these were sensations that can never be forgotten.[49]

Chapter 11

In Princely India

Each year, as the sultry hot months approached, the viceregal family fled the heat to the cool hills of Simla. With them travelled the entire machinery of state as Simla became the summer capital of British India. There was a regular cyclic pattern to these annual migrations: the summer bringing the exodus to Simla, November the return to Calcutta for the winter.

The Viceroy's train that bore them was, of course, a miniature travelling palace, consisting of a string of luxurious saloon cars painted white and gold and pulled by a gleaming black locomotive. The private car of the vicereine had her bedroom, a boudoir, a bathroom with a bathtub, a dressing room and a room for two of her maids. The Viceroy also had a study furnished with a desk and chairs.

Elsewhere in the train there was an entire car of rooms for the children. A wooden panelled lounge car with deep rugs, fans, ornate wall lights, damask-covered chair and sofas served as the gathering place for the Viceroy, the ADCs and other accompanying officers, each of whom had an apartment. One car was devoted entirely to dining. It included a kitchen equipped with stoves, ovens, hot-boxes, an ice-box, a wine cabinet and a dining room with a long table for formal meals. To see that everything moved with clockwork precision, the general traffic manager travelled on the train.[1] At every major station, there was an official guard of honour awaiting the Viceroy.

In spring and autumn, the Viceroy toured the princely states. Exotic, flamboyant, some enlightened, some decadent, the tales of wealth and peccadilloes of the Indian princes who ruled over two-fifths of India has always been legendary and somewhat reminiscent of the *Arabian Nights*. Apart from their palaces in India, they frequently maintained houses in London, Paris and the Riviera, the less fortunate reconciling themselves to reserving vast suites at hotels such as Claridges and The Savoy in London, and keeping them filled with massive bunches of freshly-cut hot-house flowers for the entire length of their stay. The search for sensation could be bizarre. One prince employed live human infants as tiger-baits at his shoots, boasting that he always got the tiger before it got the bait. Yet another prince who kept 60,000 pigeons as a hobby, once arranged a marriage ceremony for one of them complete with a banquet and fireworks, only to lose interest when the palace cat ran away with the bridegroom![2]

Enjoying treaty relations with the British Crown, the princes accepted the Viceroy's paramountcy. He represented the British Crown in India and exercised authority through the Foreign and Political Departments of the Government of India. Retaining princely goodwill had become a part of British policy after the Mutiny, for in Canning's words, these 'few patches of native government' had proved break-waters in the storm'. The princes enjoyed gun-salutes according to their status, their rank being indicated by the number—21 being the highest. A little black book kept at Government House advised the Viceroy on the honour due to a visiting prince. Accordingly, the Viceroy either descended from the steps of his throne to greet his visitor, or awaited him on the dais. Few ceremonials could have been more calculated to appeal to Curzon.

Being treated like a ruling sovereign made Curzon almost believe he was one. Once in England, the Queen cut down on the magnificence of her ADCs' uniforms; this automatically entailed a corresponding reduction in the grandeur of the Viceroy's entourage. But within his very first month in India Curzon was writing off a letter of protest to the Queen's Principal Private Secretary, Sir Arthur Bigge (later Lord Stamfordham). Overruling Curzon, Bigge had scathingly said, 'The Queen fully realises what you said...but H. M. does not feel sure if it would be advisable to have different uniforms for her ADC to those of the Viceroy.'[3]

Nevertheless, Curzon refused to be squashed. The somewhat archaic world of princely India ideally suited his temperament, being one of the last arenas where he could play out the role of a benevolent despot. 'I was delighted with Kathiawar,' he said to Mary. 'There is a flavour about it of an old time, semi-feudal society, which has crystallized into a new shape under British protection...'[4]

India was indeed the land of dreams, where the white man could play out his fantasies of grandeur and enjoy homage almost bordering on divine worship. Curzon thrilled to it, seeking out occasions to inspire, even at great discomfort to himself. Once, while touring a plague area, he manfully submitted himself to an inoculation so as to set an example. He ruefully said, 'They pump into your arm the best part of a wine-glass of disgusting fluid, which inflames the whole limb, gives you fever, causes you acute agony for twenty-four hours, and, in some cases leaves you miserably seedy for four or five days.'[5]

In fact, the Viceroy did much more. Dutifully he toured the disease affected areas, not squirming at visiting hospitals and relief kitchens in the heart of the cholera camp of Ahmednagar and Nasik, treading ground few Englishmen cared to venture. An alarmed Mary told him, 'I have been absolutely miserable over the accounts of your doings in hospitals, and Colonel Fenn [the doctor] shares my horror and anxiety.'[6]

Curzon loved the tributes inscribed on triumphal arches erected to welcome the visiting Viceroy. The fantastically adorned greetings so delighted him that years later in England he would spend hours chuckling over them. At Jaipur he was able to smile at the accidental jumbling which converted 'A Gala Day' into 'A Gal a Day'.[7] He was so enormously pleased by the sentiment that the composition of the words did not bother him for he believed his people were using their smattering of English to glorify the Viceroy.

A viceregal visit sent the state's rulers into a whirl of activity. Trains were rescheduled and vast spaces cleared to put up tent cities. Tigers were rounded up for the viceregal shoot. Palaces were overhauled, quantities of expensive furniture imported and, on one occasion, even a silver bath fitted to pamper the visiting Viceroy. Mary wrote:

> ...preparations for the Viceroy's camp entail such hundreds of coolies and mules, and a general turning upside down of the whole country. Miles of path are cut on any hill the great Sahib may walk on, and underbush cut down and a grand clean-up made of the country. This is done in each small state we go through, and as crowds follow us everywhere every bird and beast flies before this invading army.[8]

Tours of the princely states began with scarlet-bound programmes being distributed by the Military Secretary beforehand, outlining every detail of the journey. At the station, a crimson carpet would be laid from the precise spot where the viceregal carriage door would open. It stretched all the way to the carriage waiting to escort them into the city. The arrival of the train would be a signal for the guns to thunder a royal salute.[9]

The tours must have taken their physical toll. The long and wearisome journeys often left them ill and tired and the letters home were strewn with references to them. Mary was once driven to exclaim, after one particularly exhausting visit:

> I am so glad that our tour is coming to an end. The strain of it is very great and our days are filled with politics,—philanthropy and charity, and our evenings with society ... oh! the fatigue of night after night of dinner parties, frightful music & worse food, and company with whom you have nothing in common. It is all I can do to fill my part. We have been surrounded by plague and cholera,

but every precaution is taken for us. George goes through
agony with his back, but only I know it.[10]

But they were also exhilarated by the heady adulation they
encountered wherever they went. They were, after all, the
most honoured guests in each of the princely capitals they
visited. The princes indulged and humoured them and they in
turn wallowed in the adulation. It was not unnatural. The
Viceroy to them was as Victoria was to her nobility; his word
was law. The Curzons were made to feel as though they were
the most regal of a royal party. The dizzy heights of the gold
and silver *howdahs* which carried them was a far cry from the
horseback rides at Kedleston Hall. The son of a Derbyshire
country squire and the daughter of an American real-estate
dealer had indeed come a long way and there was considerable
truth in Lord Beaverbrook's observation: '... for all the rest of
his life Curzon was influenced by his sudden journey to heaven
at the age of 39 and then by his return to earth seven years later,
for the remainder of his mortal existence.'[11]

In the lush green state of Travancore, at the southern tip of
India, they cruised along the beautiful tropical backwaters, the
Maharaja having provided them with a boat shaped '... like a
Lord Mayor's barge,' with, 'a cabin, a roof to sit on, and
twenty rowers'.[12] Their entourage was however so great that,
'we needed 40 boats, which made a procession a mile long'.[13] In
Hyderabad they had been met at the station by the Nizam himself. Having heard of his repute as one of the richest men in the
world, Mary was taken aback to find a '... shy little man, who
seldom speaks,' and one 'without any of the gorgeous orna-
ments usually inseparable from Oriental majesty.'[14] A 162-
carat diamond was discovered hidden in the toe of his slipper
by his son after his death.[15] The Nizam had entertained the
Curzons in his magnificent banquet hall, its gilded ceiling supporting massive Levantine chandeliers.

But for Mary the visit had been frustrating and uncomfortable. At the state banquet the imperturbable Nizam had sat by
her side. She had racked her head to make conversation, going
through the alphabet for subjects: 'A for Arab horses, B for

Colonel Barr [the British Resident], . . . but each topic died at its birth and only produced a gentle "Yes" or "Exactly" from His Highness.' When she went on the shoot it was to witness a bizarre and tragic accident of a tiger ripping apart a *shikari's* head. At her last meal in Hyderabad, a banquet for 140 people, to her great chagrin, 'a horrid old man' sat opposite her and threw the bones of his chicken curry under the table, most of which landed on her white muslin dress. 'My wrath rose at his table manners.'[16]

At Bhopal, where 'the station was beautifully decorated and crowded with people', Mary noted with satisfaction how, as her husband got out of his railway carriage, the Begum advanced from her 'little tent arrangement on the platform' to meet him. The bands played and they were driven off along a road lined first by the Imperial Service Cavalry and later by the state army comprising of aged veterans with orange beards, holding rusty muskets. Behind them was ' . . . the most wonderful crowd of Natives, camels, elephants, in every rainbow colour; all native bands playing a kind of "God Save the Queen" and trumpets shrieking royal salutes.'[17] The show was splendid, squalid, picturesque but it was all for *them*, the Curzons.

At Gwalior things were even better. 'The state carriages were *magnificent* . . . and everything intensely spick and span . . . After all, this Maharaja Scindia is enormously rich, and was covered with jewels,' Mary noted.[18] Curzon could not have been more eloquent—calling his host 'much the most remarkable and promising of all Native Chiefs . . . he practically runs the whole state himself,' going on to add that the young Gwalior Maharaja, ' . . . in his remorseless propensity for looking into everything, and probing it to the bottom, rather reminds me of your humble servant.'[19] Coming from the Viceroy, this was praise indeed.

The Maharaja had bent over backwards to look after his viceregal guests. For his tiger shoot a perfect fairyland of a camp, complete with flowerbeds and fountains, had been erected in the thick of the jungle. A battery of elephants escorted the party to the prescribed spot. After hours of wait-

ing a tiger was seen, '...a flash of yellow dashed before
us...George shot him in the back'. But alas! the wounded
tiger disappeared into the thicket. Next morning's search also
drew a blank and the crestfallen Curzons had to return to
Gwalior empty-handed. But a few hours later, scanning out of
her palace window, Mary saw to her utter delight '...the
Maharaja in a barouche drawn by four horses, and in the
barouche two immense dead tigers! It was sight! He had gone
out and got George's wounded tiger and another, propped
them in a carriage and driven 20 miles in two hours, and there
he was triumphant and dusty with his two tigers.'[20] Interestingly, years later, Curzon was to use his Gwalior skin to win
favour with the gorgeous Mrs Elinor Glyn, celebrated on the
London stage for her bouts on a tiger skin. Apparently, upon
coming to know that the taciturn Lord Milner was wooing the
lady and had presented her with a tiger skin, Curzon had
promptly sent in his Gwalior trophy. It proved to be a winner.
Elinor Glyn shortly became his mistress.[21]

Yet, somewhat perversely, it was in the home of this 'wonderful' Maharaja, 'who is the kindest, most thoughtful host',
that Curzon decided to tweak the ears of the native chiefs. In
the banqueting hall of Gwalior, with its mirrors shimmering
across the walls and ceiling, an aghast company of princes was
made to listen to a viceregal homily they had rarely heard
before. 'The Native Chief has become, by our Policy, an integral factor in the Imperial Organization of India. He is concerned not less than the Viceroy or the Lieutenant-Governor
in the administration of the country. I claim him as my colleague and partner,' he stated. But having called them equals,
the Viceroy proceeded to lecture as though they were nothing
but errant boys:

> He cannot remain vis-à-vis of the Empire a loyal subject of Her
> Majesty the Queen Empress, and vis-à-vis of his own people a
> frivolous and irresponsible despot. He must justify and not abuse
> the authority committed to him; he must be the servant as well as
> the master of his people. He must learn that his revenues are not
> secured to him for his own selfish gratification, but for the good of

his subjects, that his internal administration is only exempt from correction in proportion as it is honest; and that his 'gaddi' is not intended to be a divan of indulgence, but the stern seat of duty. His figure should not merely be known on the polo-ground, or on the race-horse, or in the European hotel.[22]

Though considerable harm may have been done to India by those princes who did go to Europe only in pursuit of wine and women, surely there was something ridiculous in the Viceroy's schoolmasterly treatment of the princes. He certainly valued their importance to the Empire for it was on those grounds that he had asked them to a durbar. Yet he thought nothing of demolishing them publicly before the eyes of their own people.

The Chief had a duty to his people and Curzon tried his best to see that he fulfilled it. He said in a letter to the king:

> Patiala, who was in many respects a good fellow and a fine sportsman, has just died of delirium tremens, before the age of 30, the prey of English jockeys and pimps. The Rana of Dholpur, a magnificent rider, is slowly dying from the same cause. The Maharaja of Ulwar has had to be suspended for persistent sodomy. The Raja of Jind, who was under a British Officer as guardian, has just married the illegitimate daughter of the illegitimate wife of a German aeronaut. The late Nawab of Bahawalpur fell under European influence, married a European woman of low character, and died at the age of 30, a confirmed sot.[23]

There was considerable truth in what Curzon had to say and much of his criticism was justified, but he spoiled his case by his hectoring tones. Even Victoria had been driven to protest:

> The Queen-Empress quite agrees with the Viceroy that too frequent visits of the Native Princes to England is not always desirable, but she thinks that this should not be done in too peremptory a manner. It would hardly do to refuse them to come for a short time.[24]

The Viceroy was, however, in no mood to listen to strictures from the Queen. He was in the process of issuing some himself and even sent some of them to the Queen! He objected to her giving a reception at Buckingham Palace for the Maharaja of

Kapurthala whom Curzon called a third-class chief. The Viceroy created such a furore with the Secretary of State's office that an aide at the India Office was given a sharp dressing down.[25]

Persuading himself that he knew better than his monarch how to treat a native chief and to keep him in his place, Curzon could not resist telling Hamilton, Secretary of State for India so. He boasted how native chiefs would not dare to take liberties with him, citing the example of one who, while considering himself to be a favoured guest at Windsor, 'regards it as a high honour to be asked to dinner at Government House'.[26]

When Hamilton chided him for being too much of a disciplinarian with the Princes, the Viceroy retorted:

> For what are they in most part, but a set of unruly and ignorant and rather undisciplined schoolboys. What they want more than anything else is to be schooled by a firm, but not unkindly, hand; to be passed through just the sort of discipline that a boy goes through at a public school in England . . . to be weaned, even by a grandmotherly interference, from the frivolity and dissipations of their normal life.[27]

But helping the princes on the straight and honest path was also good imperial policy and Curzon was not unaware of it. He defended his stern reproofs saying:

> We do so, not so much in the interests of the Princes themselves, . . . as in the interests of the people, who are supposed to like the old traditions and dynasties and rule. But supposing we allow the confidence of the people in their rulers to be snapped; supposing we allow Native India to be governed by a horde of frivolous absentees who have lost the respect and affection of their own subjects, what justification shall we have in such a case for maintaining the Native State at all?[28]

The Coronation Durbar of January 1903 brought into critical focus Curzon's passion for the display of pomp and splendour, in which he was to play a central symbolic figure. Within a few hours of the death of Victoria, Curzon had written to King Edward VII suggesting that he come out and crown himself Emperor of India. Curzon declared:

To the East there is nothing strange, but something familiar and even sacred, in the practice that brings sovereigns into communion with their people in a ceremony of public solemnity and rejoicing after they have succeeded to their high estate.[29]

The King could not come and for a while it was mooted that the Prince of Wales might come instead. Curzon was dismayed. He was not prepared to play second fiddle. He argued:

The Viceroy represents the Sovereign. He is treated everywhere in India exactly as if he were the Sovereign; and this not by Europeans only but by all the Nobles, chiefs, and Princes of India. The appearance of no one on the scene, even though it be the heir to the throne, can deprive him of his position as representative of the Sovereign. He cannot, a Governor elsewhere, step down and take second place.[30]

Eventually, it was the King's brother, the Duke of Connaught, who came. In keeping with protocol, he was next to the Viceroy on Indian soil and he accepted his lesser position with grace.

The Coronation Durbar produced a new chapter in the history of the regalia of the Empire. Curzon with his characteristic thoroughness had supervized its organization to the last detail.[31] The Viceroy and his wife made their state entry into Delhi riding atop the largest elephant in India. As they approached, the band burst into 'God save the king' and the multitudes rose to their feet, the troops presenting arms, the European men with their topees raised and the *ryots* salaaming. In the centre of the scene sat Curzon, with right hand uplifted, as one accepting salute. Here at last, atop the glittering *howdah* with the golden umbrella, a symbol of royalty, Curzon felt he was the true British Moghul.[32] It was the supreme moment of his career. No other honour could be comparable to that which he experienced that day.

Mary added to the illusion in making her husband feel that it was he who was the chosen one. 'The Connaughts were no more than an extra gargoyle . . .' she told him. 'Neither of them have personalities—he has a most exquisite manner but she is only a German *housefrau*.'[33] Instead of recognizing her husband's tragic flaw, Mary prodded him into overreaching himself.

Six months before, the indulgent Salisbury had retired from the prime ministership and his nephew Arthur Balfour had stepped into his place. Tall, languid, donnish-looking, intellectual, and a landowner, Balfour came from a background similar to Curzon's: he had also moved from his ancestral estate to Eton and Oxford before taking his seat in Parliament. Along with Curzon, Balfour had lorded it over London society and The Souls. Curzon had always paid affectionte deference to the older man, but that did not stop Balfour from now playing deadly politics against his friend.[34]

The Durbar of 1903 formed the climax of Curzon's career as an eastern potentate. A shadow, no doubt, had been cast over the Durbar by the Home Government's refusal to accede to Curzon's request for the announcement of a remission of the salt tax, which he felt would be in keeping with the accession of a sovereign in Indian traditions. To the masses of India, the king they accepted was the king they saw delivering the largess. Interestingly, Gopal Krishna Gokhale at his presidential address to the Congress in 1905 was to reveal how, 'Four years ago, when with a surplus of seven crores or nearly five million sterling in hand, the Government of India did not remit any taxation and I ventured to complain of this in the Council and to urge an immediate reduction of the salt duty, I well remember how Lord Curzon sneered at those who "talked glibly" of the burdens of the masses and of the necessity of lowering the salt tax as a measure of relief!'[35]

Hamilton had argued that such an announcement would set off an 'awkward precedent':

> A reduction of taxation and the accession of a new Sovereign, though in accord with Eastern ideas, would establish a most awkward precedent; a similar benefit would be expected at the commencement of every reign, and unpopularity be caused if the expectation was not fulfilled.[36]

But Curzon felt he knew what was right and said so. 'These,

as I have before told you, are my views, and no abstract reasoning in the world could convince me that they are not soundest and most statesmanlike advice that is open to me to give you.'[37] When the Cabinet remained adamant, Curzon petulantly threatened not to hold the Durbar. 'Well, I say frankly that I would sooner not hold the Durbar at all than hold it under the conditions which you desire to prescribe for me. . .'[38]

The Cabinet, perhaps resentful of the power Curzon had begun to wield in India, refused permission. Angry at being thwarted, Curzon had gone over the Cabinet to wire directly to Sir Francis Knollys, the King's Private Secretary, asking for royal intervention in the matter. He also threatened resignation if he did not have his way.[39]

Curzon complained to Balfour, the Prime Minister:

I have served you well out here for four years. I have sacrificed everything in that time—health, ease, leisure, and very often popularity—for the sake of the duty imposed upon me. And now you go and ask me at the end of my fourth year to throw away the results of all this labour and devotion, and to injure the cause, *viz*, that of binding the Indian people to the British throne which is dearer to me than my life, by thrusting upon me the duty of announcing this great disappointment to the Indian people; and you do it for no reason of which I have yet been informed or can imagine. Is this fair? Is it generous? Is it just? You have never served your country in foreign parts. For your own sake I hope you never may. English Governments have always had the reputation of breaking hearts of their pro-consuls from Warren Hastings to Bartle Frere.[40]

Curzon had concluded his letter with another threat. He said, 'If a public servant has lost the confidence of his master, they have the right to recall him and you can exercise that right in the present case.'[41] But the Cabinet had refused to budge. Balfour also chided his old friend:

I cannot really assent to your view that, because the position of the Sovereign was (in your view) affected by the course to be taken at the Durbar in reference to taxation, you were therefore justified in carrying on an independent correspondence with him on a point of high policy without the knowledge or assent of your colleagues. You seem to think that you are injured whenever you do not get exactly your own way.[42]

School master Wolley Dod had made a similar complaint about Curzon to Lord Scarsdale at Eton thirty years earlier.

With the Durbar, therefore, Curzon's relations with the Home Government had steadily gone from bad to worse, with the correspondence between the two becoming increasingly acrimonious at the time of Curzon's clash with Kitchener. In 1905, when Curzon was forced to quit India, he felt it was because the British Cabinet sided with the Commander-in-Chief against the Viceroy.

In September 1903, Balfour had brought in Brodrick to succeed the retiring George Hamilton as Secretary of State for India. Though Curzon and Brodrick had been close friends for thirty years, the Brodrick that took over at India Office was not the admiring friend Curzon had known earlier. It took Curzon a long time to realise this fact. With all his gifts, nature had denied him sensitivity. Many years later Curzon was to write with corroding bitterness:

> St. John Brodrick was a greater friend of mine at Balliol and in after life than at Eton. He was in some respects my closest friend in public until an evil hour he became Secretary of State while I was Viceroy. In two years he succeeded in entirely destroying both my affection and my confidence. Burning to distinguish himself as the real ruler of India, as distinct from the Viceroy, . . . he rendered my period of service under him one of incessant irritation and pain and finally drove me to resignation.[43]

Chapter 12

An Imperialist and a Gentleman

Curzon was in Simla during his first summer when news came to him of the Rangoon outrage. Some officers of the West Kent Regiment had raped a Burmese woman in open daylight. The incident took place at the beginning of April but Curzon came to hear about it only in June, when reading a newspaper. The case had been hushed up. It would never have been investigated had Curzon not intervened.

Outraged, the Viceroy said when he mentioned it to the then Commander-in-Chief [Sir W. Lockhart] nearly two months after it had happened, he was told 'it had never so much as been reported to him.'[1] The Lieutenant-Governor of Burma had also remained silent. 'That such gross outrages should occur in the first place in a country under British rule; and then that everybody, commanding officers, officials, juries, departments should conspire to screen the guilty, is,' he cried out angrily, 'a black and permanent blot upon the British name.'[2]

Forcing the Commander-in-Chief to take strong action, the Viceroy had the offenders dismissed from the army and some senior officers relieved of their command. He then had the entire regiment banished from Rangoon, 'to the worst spot that I could find; they were accordingly sent to Aden.'³

Curzon personally rebuked the senior officials who had played a role in suppressing the affair and had the order-in-council published to inform the public of the government's anger over the matter. The Viceroy's decision was hailed by the Indian press. The British community in India was appalled. As Curzon told his wife: 'The *Pioneer* ... said not one word about the soldiers' part, *and left that out altogether from its report.* Isn't the whole thing mean?'⁴ But even Members of the Viceroy's council had not whole-heartedly endorsed Curzon's measure and the Commander-in-Chief and the Military Member warned Curzon of another mutiny.⁵ His friends felt that the Viceroy had gone too far and that his actions were 'galling to [European] national pride.'⁶

Following the Mutiny, few cases of soldiers beating coolies had been reported and fewer still brought to book. Curzon noted that 'when I came to this country, I found that in spite of excellent pronouncements on the part of many of my predecessors, the number of cases of violent collision between Europeans and natives was increasing with a rapidity that appeared to me to be dangerous and menacing.'⁷ He was disturbed, he said, that by 'the general temper and inclination of Europeans, as illustrated by the attitudes of many of our officers both military and civil, by the tone of the English newspapers, by the verdicts of juries, and by any other test that it was possible to apply, was in favour of glossing over, palliating rather than exposing and punishing these crimes.'⁸ Curzon's inquiries revealed that in the previous twenty years, among the recorded cases of violence, 84 Indians had been killed compared with only five Europeans and 57 Indians had been seriously injured as compared with 15 Europeans.

The Mutiny had sharply reversed the early faith in the equality and perfectability of mankind. Indians were now seen as being chronically deceitful and indolent, their vices racially inherent and therefore incurable. Born subservient, they were considered to be incapable of leadership. It was, therefore, the white man's burden to take them in hand. This racial arrogance indirectly owed something to Charles Darwin, whose *Origin of the Species* was published two years after the Mutiny. By laying stress on the importance of hereditary factors, Darwin had inadvertently hardened racial attitudes.

There had been a time when a warm *bonhomie* existed between the ruler and the ruled. One poet wrote nostalgically:

> When Thomson ruled in Thomsonpore
> Somewhere round eighteen eighty,
> From end to end his district wore
> An aspect warmly matey,
> No deep division severed them
> The Powers that Be from other men;
> But all was friendly to the core
> When Thomson ruled in Thomsonpore
>
> When out on tour, a word or jest
> He'd have for every ryot.
> And oft by kindly souls be pressed
> To taste their humble diet;
> Or Mr T, with playful spank,
> Would bathe the children in the tank
> And all would laugh with merry roar
> When Thomson ruled in Thomsonpore.[9]

Earlier Empire-builders could hardly escape close acquaintance with India. Being isolated in far-flung stations they did not see their fellow countrymen for weeks at a stretch and so made the best of their isolation in local company. But times were changing. Even before the Mutiny, a British resident sadly wrote home to his family noting, 'I have been saying for years that if a man who left this country thirty years ago were to visit it he would scarcely credit the changes he would universally witness in the treatment of the natives, high and low ... restraint is cast away ... and [they display] a supercilious arrogance and contempt of the people.'[10]

The mushrooming of all-white clubs with their cool interiors further drew apart the rulers from the ruled. In these shady bowers, sealed from the heat and dust outside, the only reminders of India were the silent, soft-footed, salaaming servants. The atmosphere was conductive to the sprouting of easy intimacy, and under the club's high ceilings, official and non-official British India moved closer together. Rabindranath Tagore lamented the role played by these all-white clubs in beguiling the British official away from India. 'In debate or business conflicting opinion may arouse the excitement necessary to maintain one's own point of view, but it is impossible to resist the silent or half-spoken view expressed in sport, in picnic, in female voices, and in the glances of women,' he said.[11]

No exception was made even for the Indian member of the ICS. He found himself debarred from British society though he may have 'practically cut himself off from his parent society, and lived and moved and had his being in the atmosphere so beloved to his British colleagues.'[12] Getting into the ICS had meant becoming 'completely estranged' from the society of his own people and becoming 'socially and morally a pariah' among them.[13] Surendranath Banerjea tells of the grim prospect of social ostracism that clouded his journey to England in 1868 to sit for the ICS. All preparations for departure had to be made in greatest secrecy as though the candidate was 'engaged in some nefarious plot of which the world should know nothing'. Banerjea recalled how the fact was concealed even from his mother and when at last the news was broken to her she fainted away.[14] Banerjea, along with his two companions, Romesh Chandra Dutt and Behari Lal Gupta had to spend the night before departure at the house of an eminent barrister nicknamed 'Protector of Indian Emigrants Proceeding to Europe', and start for their steamer before daybreak![15]

Upon his return home, the Indian member of the ICS made an all-out attempt to live like an Englishman. Bipin Chandra Pal recalls the time when Banerjea, now a members of the ICS and Assistant Magistrate of Sylhet, visited their school in 1871:'...I remember how one afternoon in April or May, the young Bengalee civilian came to see our school and in the course of his visit came to our classroom and "inspected" our work for a few minutes...he was dressed up as a "pucca Saheb" and talked to us in English. I have, however, more vivid mental

1 Portrait of Curzon, 1864, aged five years old. He cried when his curls were cut off.

2 With Master Archibald Dunbar, 1872. He was a 'great favourite' of Dunbar. Curzon attracted the attention of several masters who were known to have homosexual tendencies.

3 Curzon's study at Eton, with picture of his mother Lady Blanche Scarsdale above the mantlepiece. She helped to decorate the room. To Curzon she was 'very, very precious'.

4 Curzon's father, Revd Alfred, 4th Baron Scarsdale, 1909. Portrait by William Logsdail.

5 Lord Scarsdale playing host to Curzon's Southport Conservatives at Kedleston Hall, 1890. Yet an impression was built that his parents were indifferent to him.

6 Newly-wed George and Mary being feted by Lord Scarsdale at Kedleston, 1895.

7 At a country house party with the Prince of Wales. Mary is seated to his left.

8 Mary as Vicereine. 'It takes my breath away,' she said.

9 Curzon as Viceroy. Portrait by William Logsdail.

10 Government House, Calcutta. Its resemblance to Kedleston made Curzon believe
that destiny had chosen him to rule India.

11 9th Earl of Elgin, Viceroy of Indi
1894–9. The short, bearded Elgin wa
dwarfed by his statuesque successor.

12 Bal Gangadhar Tilak with a picture of
Maratha hero Shivaji in the background.
Arrested in 1897 for inciting murder of
plague officer Charles Rand, Curzon
had ordered his early release.

13 Moderate Gopal Krishna Gokhale in western dress, seated on extreme left. As
President of Congress in 1905 Gokhale compared the Curzon administration
to that of Aurangzeb saying 'there we find the same . . . policy of distrust
and repression'.

14   First session of the Indian National Congress, Bombay, 1885.

15   Bankim Chandra Chatterjee whose poem '*Bande Mataram*' became the anthem of militant nationalism.

16   Surendra Nath Banerjea. Dismissed from the ICS for a technical offence in 1875, he was twice president of the Congress.

17 Allan Octavius Hume, ICS, General Secretary of the Indian National Congress, 1885–92. Hume was responsible for killing more than a hundred rebels during the Mutiny.

18 Bipin Chandra Pal, who founded the extremist newspaper *Bande Mataram* along with Aurobindo.

19　Aurobindo's study at Cambridge. Compare its stark interior to the opulence of Curzon's study at Eton. Aurbindo was disqualified from the ICS for failing to appear for a riding test. The partition of Bengal gave him the opportunity to become the high priest of militant nationalism.

20　Aurobindo at school in Manchester, 1881.

21　Aurobindo with his wife Mrinalini Devi at Baroda.

ir Andrew Fraser,
Lieutenant-Governor of
Bengal, 1903–8, one of the chief
architects of the Bengal
partition. Bust by George
Frampton, 1911.

23 Sir Bampfylde Fuller, first
Lieutenant-Governor of the new
province of Eastern Bengal and
Assam, he antagonized Hindus by
likening Muslims to a favourite wife.

24 Nawab Sallimullah of Dacca to whom
the Government of India sanctioned a
loan of £100,000.

25 Old Court House Street, Calcutta, 1882.

26 Palace of Jaykrishna Mukherjee. Built in 1840 it illustrates the way of life of a zamindar in nineteenth century Bengal.

27 The viceregal chairs at Government House, Calcutta.

...rrakpore House: country residence of the Viceroy.

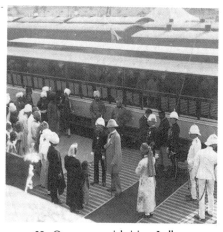

29 On a ceremonial visit to Jodhpur.

30 On a tiger shoot.

31 In princely India: ascending the palace steps with the Maharaja of Orchha.

32 The Viceroy making a state entry at the Delhi Durbar, 1903, playing out fantasies of being the Great White Moghul.

33 On an archeological survey, Mandu. Pandit Jawaharlal Nehru said, 'Curzon will be remembered because he restored all that was beautiful in India'.

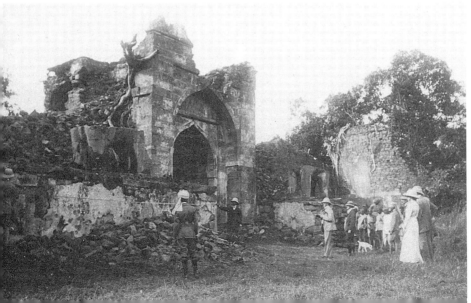

34 The Viceroy with his young family, 1903. Tales of their 'foolish formality and stiffness' had by then begun to do the rounds.

35 Mary on arrival in India as Vicereine. She found the British women 'chilly' and 'critical' and they huffed at having to curtsy to an American girl.

36 Lord Kitchener, Commander-in-Chief in India, 1902–9. Well-meaning friends had warned the Viceroy, 'he is... unscrupulous in his methods'.

37 St. John Brodrick. A close friend for thirty years, but the Brodrick who took over as Secretary of State for India, 1903, was not the admiring friend of earlier times.

38 Mrs Elinor Glyn, actress, novelist, and Curzon's mistress for eight years. Curzon had courted her with a tiger skin. Portrait by Philip de Laszlo.

39 Mrs Grace Duggan who married Curzon in 1916. When asked to go to Kedleston, she said, 'I would much rather not go at all—after all one's home is where one's heart is'. Portrait by Philip de Laszlo.

40   In the robes of Chancellor of Oxford University, 1907.
Portrait by Sir Hubert Von Herkomer.

41 Curzon's Taj Mahal: the Victoria Memorial, Calcutta. Fashioned in the
'Italian Renaissance' style it was unable to escape the Moghul touch.
Interestingly, the original pedestal for Curzon's statue in the forecourt of
the Memorial is now occupied by a statue of Aurobindo.

pictures of Mrs Banerjea, who, riding, after the fashion of those days, on a high pony, with flowing skirts and veil that covered her hat, was the wonder of the town.'[16] Despite these attempts, the British officials at Sylhet 'did not freely accept the young Bengalee and his wife, who had forced himself into their preserve, into their sociey'.[17] The only friend the Banerjeas had was an Armenian Deputy Magistrate, 'whose clothes, though not his blood or colour, lent him something of the dignity of the ruling caste'.[18] It was this Armenian who found time 'to accompany Mrs Banerjea in her rides through the town'.[19]

Racial abuse was far worse in the army, few eyebrows ever being raised when soldiers returned drunk from a party and beat up their servants. On tea gardens, the coolies had long learnt to take kicks and slaps as a part of their daily routine. Enlisted from districts of upper India by a force of paid recruiters of British business houses in Calcutta, the coolie's plight was comparable to that of a slave. To cover the cost of transporting him to Assam, the law authorized his engagement under a five-years' indenture, giving plantation managers the extraodinary right of arresting an indentured man if he absconded![20]

A visitor to a tea garden recalled how 'on some gardens the coolies were virtually prisoners' and how he came across notices posted at river ferries and railway stations describing runaway coolies which '...reminded one of *Uncle Tom's Cabin*'. The visitor added, 'On some gardens there was a good deal of flogging,' and remembered an occasion when 'in one of them a woman was stripped and flogged. Her husband brought a criminal charge against the garden overseer. He was acquitted by an Indian Assistant Magistrate on the score that he acted under the orders of his European manager. No further action was taken.'[21]

Ripon's Ilbert Bill brought racial animosities to a head when it became 'the signal for a tearing agitation among the European residents in the country, who were literally maddened by his attack on what they believed to be one of their fundamental rights as European British subjects.'[22] The then Lieutenant-Governor of Bengal, Sir Rivers Thompson, openly gave vent to his racial prejudice:

> The very bad thing about the bill is its principle—the principle, that is, by a stroke of the pen we are to establish equality; ignoring race distinctions, . . . Our thoughts are not their thoughts nor are their ways our ways; and it has been quite justly pointed out that as long as there is such widespread divergence between Englishmen and Natives, as regards moral standards, social customs and political status, any attempt to remove judicial disqualifications must be as dangerous as it is premature.[23]

Old hatreds and fears came to the fore. Mr Branson, a member of the Calcutta High Court Bar, in a public speech in Dacca made 'a most savage attack upon Indian culture and character citing . . . medieval social institutions of caste, child-marriage, *zenana* seclusion and the prohibition of widow remarriage as conclusive evidence' of Indian moral degeneration, which 'branded' Indians with 'absolute disqualification to sit in judgement upon European criminals'.[24]

The young Viceroy was not unaware of these simmering fires and said, 'I am not so foolish as to ignore the sentiment of racial prejudice'.[25] But nobody—least of all his own people— was going to browbeat him into withdrawing from his mission 'to fight for the right, to abhor the imperfect, the unjust and the mean, to swerve neither to the right hand nor to the left, to care nothing for flattery, applause or odium or abuse.'[26] To convince the mass of people that British rule is 'juster, more beneficent' was a part of the benevolent despotism that Curzon had envisaged for himself in India. At such times even his inherent racialism took a back-seat. Indeed, he was soon gathering for himself a reputation for 'fearlessness, omniscience, and justice and action', and the Secretary of State admiringly complimented him for having placed 'the whole foundations of our Government on a stabler and more sympathetic basis than before'.[27]

In December 1902, a coolie on a tea garden in Assam was flogged to death with a stirrup leather by a young European assistant called Bain.[28] The victim suffered from an enlarged spleen which ruptured under the merciless blows. Bain was tried, found guilty and sentenced to eighteen months of simple imprisonment. Sir Bampfylde Fuller, Chief Commissioner of Assam, while believing the punishment to be light, tried to persuade the Viceroy to let it stand as he felt it left the feelings of the British planting community comparatively undisturbed.

Curzon, in no mood to be persuaded, promptly ordered that an appeal be made to the High Court for an enhancement of the sentence. The Viceroy's order infuriated the European business community of Calcutta and the judge who heard the appeal actually acquitted Bain altogether. Sir Bampfylde said, 'I felt very sorry for the youth. Beyond doubt he had acted under orders of his garden manager, and had refused to give him away—a married man with children to whom conviction would have been ruin.'[29] The Chief Commissioner went out of his way to see that the English police officer escorting Bain to prison in Calcutta would make the thing as smooth as possible for him.

But perhaps the most shocking case was that of the celebrated cavalry regiment, the 9th Lancers. Drawn from the British aristocracy, the 9th Lancers had high and powerful connections back home. In their gold-laced uniforms, their chests emblazoned with medals won in the South African War, they presented a resplendent sight.

Late in the afternoon of 9 April 1902 the regiment arrived in Sialkot from South Africa. There was general drinking all round and around 9.30 that evening three of the troopers asked their native cook Atu to provide them with a woman. When he refused, they savagely beat him up. Breaking two of his ribs, they bunged his eyes and rained blows on his head, face, back and chest. Atu was then left all night, bleeding and unconcious. The next morning he had to be carried on a stretcher to a hospital. When the District Superintendent of Police and inspector visited the scene they found the ground stained with crimson blotches of blood. Nine days later Atu died.

'No attempt had been made by regimental authorities to investigate the case,' said the shocked Viceroy, the news once again coming to his ear only when the dead man's relatives wrote to him. Apparently a court of inquiry had been held. It was a sham, Curzon angrily said, 'It consists of two Captains and a subaltern, not one of whom, so far as I can make out, understand a word of Hindustani. Their idea of taking evidence and holding an inquiry consists in examining four witnesses, all natives. They never think of sending for the doctor or for a non-commissioned officer or a soldier or for the inspector of police. In fact they make not the smallest effort to arrive at truth...'[30] Fuming with rage, he added, 'I will not be a party to any scandalous hushings up of bad cases of which there is too much in this country, or to the theory that a white man may kick or batter a black man to death with impunity because he is only a d...d nigger.'[31]

Believing this travesty of justice a crushing blow to British rule, the Viceroy hustled the Commander-in-Chief into action. Adjutant-General Boyce Combe was despatched to Sialkot and Lieutenant-General Commanding Sir Bindon Blood ordered to oversee a fresh inquiry. But resenting the Viceroy's rebuke, the sullen Boyce Combe made out 'the best case for the regiment that he could'. What really shocked the Viceroy, however, was the attitude of Sir Bindon Blood. In his determination to condemn the dead Atu, Sir Bindon did his best to discredit him, denouncing him as a drunkard and a liar, even trying to put the blame for his death on another regiment. 'Disciplinary action is quite out of question,' said Sir Bindon, recommending mere extra sentry duty as the only punishment to a regiment which had two murderers in its ranks.[32]

The infuriated Viceroy overruled the general and had the whole regiment punished: all officers and men on leave in India were recalled, and no more leave granted for six months. Ruminating on the consequence of his action, Curzon said:

> If it be known that the Viceroy, backed by the Secretary of State, will stand up even against the crack regiment of the British Army, packed though it be with dukes sons, earls sons and so on—then a most salutary lesson will be taught to the army. If we yield to military and aristocratic clamour no Viceroy will dare to go on with the work that I have begun.[33]

In some alarm, the King protested at the Viceroy's stand, which he thought unduly harsh. London gossip had it that Atu had not really been assaulted by a member of the 9th Lancers and that the Viceroy had high-handedly overruled the military authorities in meting out the sentence.[34] But Curzon stuck to his guns saying he would not sacrifice 'what I regard as the most solemn obligation imposed upon the British race to the licence of even the finest regiment in the British Army'.[35]

The 9th Lancers, as expected, took every chance to take their revenge against the Viceroy who dared to publicly cast aspersions on their famous name. Ironically, it was Curzon who gave them the biggest opportunity to do so. The 9th Lancers had been scheduled to take part in the Delhi Durbar of January 1903, which marked the zenith of Curzon's viceroyalty. With the changed circumstances, the Military Secretary advised the Viceroy to withdraw them. Curzon, in his characteristically haughty manner, refused to do so. In a communiqué to the press he boasted that he had rejected the Military authorities' proposal to strike them off the strength of troops to attend the Durbar, 'because of the illustrious record of the regiment and the desire to spare it a public disgrace'.[36]

It was a grave mistake. At the Durbar, the 9th Lancers rode by in all their faultless grace, with gleaming helmets and prancing steeds, presenting a magnificent spectacle. The European populace, as though unable to contain themselves at this dazzling sight, surged to their feet giving the regiment an hysterical ovation. To Curzon's horror, even his guests joined in the cheering. The Duke and Duchess of Connaught, who had espoused the cause of the 9th Lancers, could not conceal their pleasure at what they felt was a public vindication. Mary resentfully perceived 'the flash of glory' that the regiment's action gave to the royal visitors.[37] Sir Bindon Blood was beside himself with delight. Years later, writing in retirement in England, Sir Bindon took out his wrath on Curzon for having dared to overrule his judgement. The former Army Commander expressed spiteful malice at the European guests having publicly humiliated the Viceroy.[38] Curzon only moaned, 'I could do nothing but deplore my own ill-timed generosity in ever having allowed the regiment to go to Delhi at all.' [39] But he had no regrets over his disciplinary action. Downcast with shame over the behaviour of his people, he took satisfaction in being able to do what he felt was right:

> The 9th Lancers rode by amid a storm of cheering. I say nothing of the bad taste of the demonstration. On such an occasion and before such a crowd (for of course every European in India is on the side of the Army in the matter) nothing better could be expected. But as I sat alone and unmoved on my horse, conscious of the implication of cheers, I could not help being struck by the irony of the situation. There rode before me a long line of men, in whose ranks were most certainly two murderers, amidst the plaudits of their fellow countrymen. It fell to the Viceroy, who is credited by the public with the sole responsibility of their punishment, to receive their salute. I do not suppose anybody in that vast crowd was less disturbed by the demonstration than myself. On the contrary, I felt a certain gloomy pride in having dared to do the right.[40]

In that one supreme moment of his viceroyalty Curzon came closest to touching the hearts of the Indian people who also watched the white man's glee with angry silence.

The Army never forgave the Viceroy. 'He at once antagonised as he must have known was inevitable, his own caste in England, from which the officers of the regiment were drawn, and the whole of the British Army society in India,' observed a British civilian.[41] But he had neither flinched nor been cowed by the threats as had his Liberal predecessor, Lord Ripon. Denounced as 'nigger-lover' by sections of the European community, Curzon was held responsible for what they felt was a growing arrogance among the natives. Utterly unafraid, the Viceroy told the Secretary of State that there was little chance of improvement until a British soldier was hanged for murder![42]

No account of Curzon, therefore, can be complete without taking into account his supreme moral courage. Having taken a position, he did not once draw back: demonstrating that in India he could be both an imperialist and a gentleman.

Nor should his contribution to the architectural heritage of India be ignored. Though the Archaeological Survey of India owed its inception to Curzon, he was not grudging in acknowledging the contributions of his predecessors. He gave credit to Lord Minto for having appointed a Committee to conduct repairs of the Taj Mahal, to Lord Hastings at Fatehpur Sikri and Sikandara, and to Sir John Lawrence for saving the Jama Masjid at Delhi in the aftermath of the Mutiny.[43] Deeply pained by the neglect Indian monuments had suffered, Curzon reported to Godley, the Permanent Under-Secretary of State for India:

> In the past we have scandalously neglected this duty, and are now only tardily awakening to it. I am, therefore, personally going round with the archaeological director of the province not merely every one of the principal monuments, but every nook and cranny of every one of them, and am giving orders as to what is to be done. I do this at every place I visit, and I hope at the end of five years to have effected a very perceptible change.[44]

He did. Shedding aside his usual racial arrogance, Curzon brought a sense of reverence to India's past:

> If there be anyone who says to me that there is no duty devolving upon a Christian Government to preserve the monuments of a pagan art, or the sanctuaries of an alien faith, I cannot pause to argue with such a man. Art, and beauty, and the reverence of that is owing to all that has evoked human genius or has inspired human faith, are independent of creeds, and in so far as they turn on the sphere of religion, are embraced by the common religion of all mankind. Viewed from this standpoint, the rock temple of the Brahman stands on precisely the same footing as the Buddhist Vihara and the Mohammedan Masjid as the Christian Cathedral.[45]

In Lahore, Curzon took over the exquisite Moti Masjid created by Jehangir but then being used with callousness as a government treasury. The arches had been filled with brickwork, and the marble floor excavated as a cellar for the reception of the iron-bound chests of rupees. 'Ranjit Singh cared nothing for the taste or the trophies of his Mohammedan predecessors, and half a century of British military occupation, with its universal paint pot, and the exigencies of the Public Works engineer, has assisted the melancholy decline,' rued Curzon.[46]

A colleague remembered, '... I recall his untiring activities, in spite of sun and heat, of the long climbs among the ruins, and the long hours spent in recording what could be preserved. I feel now, though I did not feel it then, that his labours were not in vain... In one of the most beautiful of Moslem buildings we found a squatted Post Office, half brick, half adobe, and the Viceroy in his indignation ordered the whole staff to quit on the spot.'[47]

But it was the Taj Mahal that wooed him. Curzon had been hypnotized by its beauty when he had seen it as a young man on his first visit to India. Now he dedicated himself to restoring it and its environs to its Moghul glory.

In his zeal to provide a beautiful silver hanging lamp of Saracenic design to be hung over the tombs of Shah Jehan and his Queen, he carried on a six-month correspondence with Lord Cromer, Governor-General of Egypt. Eventually he visited Cairo and picked up the one he wanted. The silver lamp still hangs in the Taj Mahal today, a personal gift from Curzon. Curzon's dedication to India's monuments did not go unrecognized in free India. Prime Minister Jawaharlal Nehru said: 'After every other Viceroy has been forgotten, Curzon will be remembered because he restored all that was beautiful in India.'[48]

Chapter 13

The Giant Awakes

Almost immediately after taking office the new Viceroy sat down to work in the great study in the south-west wing of Government House. From here, on the verandah flanking the back of the house, the Viceroy laboured every day and well on into the night for the good of India as he perceived it.[1]

The first occasion for Curzon to demonstrate that he was a true imperialist came with the passing of the Calcutta Corporation Bill. The introduction of the elective principle in the Corporation and the enlargement of the Councils by the Act of 1892 had facilitated the emergence of articulate Bengalis in the political arena, much to the anxiety of the ruling community. Consequently, in 1897, Sir Alexander Mackenzie, Lieutenant Governor of Bengal, had introduced a bill curbing the powers of the Corporation. The Chairman was to have powers independent of the Corporation, even though the fifty elected Commissioners, representing two-thirds of the Corporation, were to retain their numerical strength. The remaining one-third was made up of fifteen Government nominees and the people representing commercial interests. The authority of the Corporation was to be further restricted by the creation of a

General or Executive Committee of twelve nominated members who were to act as a co-ordinating and independent executive authority. This Bill was, in effect, to transfer actual control of the affairs of the city to an executive committee which was official in character; the representatives of the rate-payers were to be left with little power except to debate. In short, the Mackenzie Bill aimed at officializing the Corporation.

In an address of welcome presented by the Municipal Corporation of Calcutta shortly after Curzon's arrival, reference had been made to the Bill which spelt doom for the future of municipal self-government in Calcutta. The address ended with an appeal for help to the new Viceroy. Curzon recorded:

> All parties appealed to me as an absolutely impartial and unprejudiced umpire. This appeal came equally from Calcutta, and from the House of Commons. I was asked to save Calcutta from the certain warfare of factions with which it was threatened, and from the destruction of Local Self-Government by the supersession of the elected 50 by the nominated 12.[2]

In a letter to Curzon soon after his arrival in India, Hamilton, Secretary of State for India, stated how he was told by Elgin, the former Viceroy, of the necessity for the drastic measure suggested in the Mackenzie Bill. Hamilton was inclined to be cautious. While endorsing the necessity for a strong executive with overriding powers, Hamilton wished the objective could be achieved without 'so summarily curtailing the representative element'.[3]

Curzon, however, felt that the Mackenzie Bill did not go far enough. 'He saw no disloyalty in India at the present time,' nationalist leader, Gopal Krishna Gokhale said, 'but he thought there was nothing to prevent the people becoming disloyal one day, and therefore his plan was to cripple them once and for all, and thus make them incapable of acting disloyally if ever they inclined to do it.'[4] Curzon himself seized upon the contradictions in the Bill in order to turn them into advantages. He argued that the 'arrangement by which the old Corporation of 75 members is left, with its 50 members propped up for show, like dummies against a wall, while nearly the whole of their powers and prerogatives are transferred from them to a new body constituted on a radically different basis, and in opposite constituent proportions' was wholly illogical. He said, 'There will be no good in dethroning the Baboo, and setting up the Englishman in his place, if the latter is not prepared to play his part in the new Constitution.'[5]

The Bill had reached the Select Committee stage in the Legislative Council when Curzon made his dramatic swoop. With one stroke the numerical strength of the elected rate-payers was cut to half. Moreover, their representation on the all-powerful twelve member Executive Committee was already cut down to one-third. The remaining two-thirds on the Executive were to be equally divided between four to be elected by the nominated and commercial interests and four appointed by Government. The reason given for these being Government appointees on the Executive was that they were there to give representation to minorities and thereby safeguard them.

To bypass the Indian argument, which Curzon admitted 'is not without force' that the Government of India had been careful to reduce only the Indian element while they had left the Government and European element intact, he came up with a cunning answer which he sent to the Lieutenant Governor, Sir John Woodburne:

In the existing Corporation I understand that among the Government nominees are 8 natives, as well as 1 sent up by the Chamber of Commerce, i.e. 9 in all out of the 25; viz., 8 Government nominees and 2 nominees of the Chamber of Commerce. Now why should you not graft on the existing constitution, so far as it affects these 25, a suggestion made in my original plan of a Corporation of 48, i.e. why should you not redistribute the 10 seats, now given to commerce by assigning 3 or 4 of their number to representative native bodies, say to the Mohammedan Association, British India Association, and National Chamber of Commerce? You would thereby be dispensed from the necessity of including so many natives among you as nominees; while if the British Trade Association complained of reduction of their existing representation (which I have always thought unduly large), you would be able to include any men whom they specially desired but were unable to nominate in your own fifteen.[6]

He hoped thus to gain 'a permanent European majority upon the Corporation as well as upon the Committee'.[7] What is perhaps more surprising is the note of naïvety with which he ends the letter. He hoped, 'You may perhaps in this way be able to placate native sentiment without in the least spoiling your scheme of future government.'[8] Indians were to him children to be manipulated.

W. R. Bright, Chairman of the Corporation, wrote to Curzon's Secretary, Sir Walter Lawrence, 'I have no doubt that, if Government had not retained the power of nominating one-third on the General Committee, there would certainly have been no Mohammedan, and probably no Eurasian on it.'[9] This was pure window-dressing and they both knew it.

Bright had hailed Curzon's proposals because, he said, they removed the anomaly of the Mackenzie Bill. He had felt the real danger of the latter was the friction which would have inevitably occurred between the Corporation, which had preponderance of one class, and the Executive Committee, which was dominated by a majority of the other class. 'Now the preponderance of the elected commissioners on the General Meeting has been done away with, I do not see any reason why there should be any effective friction between the two bodies.'[10]

Twenty-eight Indian Commissioners, including Surendranath Banerjea, the eloquent nationalist, resigned in protest and newspapers in Calcutta carried black borders to mark the death of the local self-government in the city. Only two weeks earlier Curzon had smugly dismissed all rumours about the threatened resignations: 'I am not in the least afraid of its being carried out in Calcutta.'[11] As always, Curzon failed to gauge the pulse of the people. Nor had Curzon seemed to realize the significance of the fact that this time it was not merely Congressmen, the traditional foes of the British, who had protested but that prominent landowners had also resigned.

Curzon's handling of the Bill throws some interesting

sidelights on his character. The Mackenzie Bill had been endorsed by the Lieutenant-Governor, sanctified by the Secretary of State and the Government of India, and already placed before the Bengal Legislative Council at the time of Curzon's arrival. Most Viceroys in his place would have hesitated to get embroiled in what had become a controversial issue, particularly at the onset of a term of office. Curzon was not unaware of this. 'If I were to pass the present Bill, which has been fathered, sanctioned and pushed forward by other hands than mine, I could scarcely be reproached or attacked.'[12]

But the energetic Viceroy could not resist the temptation to put his own stamp upon the measure. Besides, the greater the challenge, the more hazardous the task, the greater became his pride in having dared to do what he thought was right by the Empire. He realized, 'my intervention can hardly fail to bring me abuse from one side or the other, possibly both', but he believed 'the interests at stake, namely the future government of Calcutta, and the reputation of the British community here, are far too great to allow any such consideration to stand in the way.'[13]

What, however, is harder to understand is the fact that Curzon should have felt no qualms of conscience at having betrayed the corporation members who had turned to him as an impartial arbitrator. On the contrary, he took a ghoulish delight over having discomfited them. He wrote to Hamilton, 'I am bound to say having unanimously appealed to me as a thoroughly impartial arbiter, it would be somewhat ridiculous if they now turned round and attacked me for the result.'[14]

Threatened by the rise of the English-educated and politically articulate Indian intelligentsia, British imperialists turned their attention to the *ryots*. Following this well trodden path, Curzon looked to them in order to demonstrate that it was their British overlords and not the zamindar-dominated Congress who had the interests of the 'silent toiling masses' at heart. He got the opportunity he was looking for when Romesh Chandra Dutt, the civilian who became famous for his economic history of British India, addressed some open letters to the Viceroy on the question of the Indian peasantry. Sir Bampfylde Fuller had been called in to prepare a rejoinder. But the issue was complex and the Viceroy, as usual finding the efforts of his colleague inadequate, had taken it upon himself to write it out. In long-suffering tones he reported to Mary who was holidaying in England:

> I have been engaged in writing our big despatch about the question of Land Assessments in reply to Dutt and our critics. Holderness would have done it for me admirably had he been here. But Fuller, who is an execrable writer, sent me up seventeen pages of print, unintelligible to any one but an expert, and utterly hopeless from every point of view. Consequently I have had to do it myself. It is the most abstruse, technical, & difficult subject in the world, and here am I a Viceroy who has only been for two and a half years in the country having to write a great pronouncement upon it.[15]

The famine of 1899–1900 was devastating. The Viceroy toured the famine areas and even visited cholera camps. He took time off to listen to petitions and complaints. In fact, he enjoyed giving these audiences, and Hamilton said, 'You are truly the court of appeal against the decision of a close and united Corporation, namely the Indian Civil Service.'[16]

Curzon also sought to bring the zeal of undiminished purpose to the subjects of higher education in India. Standing before 'a dense throng of Hindu and Mohammedan students' at the Calcutta University Convocation within two months of his arrival, the Viceroy had read out his address, which his adoring wife reported, 'was very fine, as he spoke with great authority and decision: even those who could not understand it must have been impressed'.[17] Certainly the handsome young Viceroy had cut a dashing figure in his dark blue velvet robe. And he had not lost the opportunity to fire his first salvo at the educational system calling it 'faulty but not rotten'.[18]

A year later he was more categorical: 'to call upon the State to pay for education out of the public funds, but to divest itself of responsibility for their proper allocation to the purposes which the State had in view in giving them, is to ignore the elementary obligations for which the State itself exists.'[19]

A year later he stepped up the pressure and at the convocation address declared that the Government of India contemplated a more diligent discharge of its own responsibilities and as chancellor, he temporarily stopped the election of fellows to the Calcutta University.[20] In September he called a conference in Simla to study the problem.[21]

Ironically, long before Curzon's arrival, enlightened Indians had begun to look askance at the rapid westernization of their youth—since the time when, in the Viceroy's words, 'the cold breath of Macaulay's rhetoric had passed over the field of Indian language and text books'. Obsessed with the regenerative value of English education, Thomas Babington Macaulay, as president for the Committee of public instruction had, in 1835, persuaded the Government of India to incorporate his proposals into its educational policy. The East India Company's Education Despatch of 1854, with its special grants-in-aid, further encouraged the setting up of English-medium schools and colleges, financial aid being given irrespective of the religious leanings of the persons or the institutions concerned.

A number of colleges for education in English had thus sprung up by 1857 and Universities incorporated in Calcutta, Bombay and Madras. In 1862 Calcutta University conferred its first MA degree. In 1863 Patna College was established, followed by the government colleges in Lahore and Delhi and the Canning College in Lucknow a year later. The Poona Sarvajanik Sabha emerged in 1867, the Punjab University was incorporated in 1882 and the Deccan Education Society in 1884.

By 1871 four Indians had successfully passed into the Indian Civil Service—Satyendranath Tagore, Surendranath Banerjea, Behari Lal Gupta and Romesh Chandra Dutt—a number large enough to alarm Lord Salisbury, then Secretary of State for India into hastily lowering the age-limit from twenty-one to nineteen years, thereby making it virtually impossible for Indians to compete. Significantly, the four successful civilians were all from Bengal.

But not all Indians were inclined to view English education as an unmixed blessing. When the high-handed Lieutenant-

Governor of Bengal, Sir George Campbell withheld grants to higher educational institutions on the plea that he was diverting them to primary education, he had a sympathetic hearing from an unusual quarter. Poet and novelist Bankim Chandra Chatterjee refused to join in the widespread attack on Campbell.[22]

To ward off the 'too slavish imitation of English models', which Curzon had pointedly decried, had thus not been the white man's burden alone.

Wealthy zamindar and reformer Debendranath Tagore had started the Tattabodhini Pathsala to foster education in his mother tongue and to provide training in Hindu religious texts. His father, Sir Dwarkanath, had been among the first Indians to visit England and his son, Satyendra the first to pass into the ICS. In Debendranath the spirit of Indian culture and Hinduism rose up in arms, as it were, to assert itself against the onslaught of European Christianity and alien domination. Pledging himself to promote the vernacular, Debendranath is said to have once refused a letter from his son-in-law merely because it had been written in English. His son, Rabindranath, published his first poem in the *Tattabodhini Patrika* in 1874. Questioning the wisdom of teaching children in an alien tongue, Rabindranath started 'magic lantern' lectures and in 1901 opened his school in Bolpur near Calcutta to realize his ideal.

At the opening of the Education Conference in Simla, the Viceroy had criticized the examination method which forced students to stuff their brains with 'the abracadabra of geometry and physics and algebra and logic, until after hundreds, nay thousands, have perished by the way, the residuum who survived the successive texts emerge in the Elysian fields of the BA degree.' As Amles Tripathi has said, 'Tagore said the same thing in *Sikshar Herfer* and *Tota-Kahini*'.[23]

Wise government policy might have swum with the tide, giving the current a push when the going got slow, rather than churning up the entire river bed. But this is exactly what the restless Viceroy proceeded to do, even though in a more rational moment he had told the Secretary of State, 'we could no more avoid bringing our law and our education, that we could help, sooner or later, introducing umbrellas and kerosene-lamps.'[24]

Curzon said he deplored the idea of universities being turned into purely examining institutions and said he wished to see them develop as teaching institutions, his idea of a college being one 'with a history, a tradition, a "genius loci" a tutorial staff of its own'. Satishchandra Mukherji, editor of *Dawn* was expressing similar sentiments in 1902. An early advocate of self-reliance and a national system of education, in his essay on 'An examination into the Present System of University Education in India and a Scheme of Reforms', Mukherji, not unlike the Viceroy, pleaded for dedicated teachers who would evoke an atmosphere of creative research.[25] Curzon urged the growth of technical education, ' . . . to resuscitate our native industries, to find them new markets and to recover old, to relieve agriculture, to develop the latent resources of the soil, to reduce the rush of our youths to literary courses and pursuits, to solve the economic problem and generally to revive a Saturnian age.'

No doubt, reforms in the university senates had been sorely needed. As Curzon noted, the Senate of Allahabad University then numbered 82 members, that of Lahore 104, Calcutta 180, Madras 197 and Bombay 310, there being no proper reason for the disparity in their composition. The academic qualifications of the members were overlooked and there were fellows so illiterate that they could scarcely sign their own names.[26] The constitution and compositon of the senates and syndicates which governed the universities had often been filled by men whose interests were hardly educational. The universities had indeed reduced themselves to being little more than a collection of lecture rooms and laboratories and had become absorbed in conducting examinations. The primary aim of education had, therefore, been lost and cramming had become an art.

The need for national education based on indigenous needs was thus being voiced in India. A wiser ruler should have encouraged the rising tide instead of fighting against it, but then such a person could not have been Curzon, who would always fight the tide. The Viceroy deluded himself into believing that people would take his reforms in the idealistic light in which he presented them. But how could he expect them to see anything but a sinister imperialistic plot to thwart the English-educated and politically conscious Indian? Nationalist leaders dominated the Calcutta University Senate and through this platform managed to get themselves elected to the Bengal Legislative Council: plugging the Senate was naturally construed as one more method of choking off their voice.

Moreover, at the Educational Conference convened by the Viceroy not a single Hindu was invited to participate in the deliberations concerning their future. Had he thrown out an olive branch to the Moderates, and invited their representatives they may not have questioned the Viceroy's bona fides. Sir Gurudas Banerjee was included in the Commission only as an afterthought.[27] The irony of this was deeply felt, particularly when the Viceroy had the high-handedness to declare, 'Conceal-ment has been no part of my policy since I have been in India, and the education of the people is assuredly the last subject to which I should think of applying such a canon.'[28] S. N. Banerjea pointedly remarked:

> Never was there a greater divergence between profession and practice. And the effrontery of it lay in the emphatic denunciation of secrecy at the very time, and in connexion with the very subject, in regard to which the speaker had deliberately made up his mind to violate the canon that he had so eloquently proclaimed. But that was Lord Curzon's method, and we Orientals regarded it with a feeling of amusement, as coming from one who had extolled the ethics of the West above the baser morality of the East.[29]

The Moderates who might have gone along with the Viceroy felt outraged. Hamilton advised him to encourage and carry along the loyal 'Older India'.[30] But the Viceroy did not think it necessary to soften the blow or change his style. The imperial creed was to govern irrespective of fear or favour. Had not his mentors, Stephen and Strachey, said that an autocratic govern-ment was based not on popular acclaim but on force and should not be squeamish about declaring it? In fact, he seemed to take a sort of schoolboy's relish in having himself condemned. 'The

Town Hall and the Senate Hall of the University have been packed with shouting and perspiring graduates, and my name has been loudly hissed as the author of the doom of higher education in India,' he reported to the Secretary of State.[31] The operation of the law depended upon the ruler's coercive powers which should not be swayed by popular outcries. 'The native press lauds one to the skies as a sort of reincarnation of the Deity on Monday—a single adverse decision or unpalatable remark will cause one to be denounced as a Judas on Tuesday.'[32] These were, the Viceroy believed, the hazards of Indian administration and he stoically accepted them as such. What is more astonishing is that the Viceroy was not unaware of the growth of nationalistic feelings and he admitted, '...there does exist, and there is growing in the background a steady volume of public opinion as distinct from that of the man in the street, or what you call the half-penny press, which it is becoming increasingly difficult for the Government to ignore.'[33]

Indeed, the air was crackling with a brooding, sullen electricity. People had not taken kindly to the Viceroy's Corporation Bill and the publication of the Indian Universities' Report in June 1902 had sent out shock waves. The Bill was bitterly criticized in the press and from the public platform. Banerjea said:

> Under the plea of efficiency he [the Viceroy] had officialized the Calcutta Municipality; under the same plea he now proceeded to officialize the universities, and to bring the entire system of higher education under the control of Government. Efficiency was his watchword; popular sentiment counted for nothing, and in his mad worship of this fetish Lord Curzon set popular opinion at open defiance.[34]

When the Bill was finally passed on 21 March 1904, indignation boiled over. Rumours about an impending partition of Bengal whirred like arrows and the Universities' Act compounded the hysteria. Overnight, hundreds of students were out on the streets voicing anger and protest.

Bal Gangadhar Tilak's arrest in the summer of 1897 had fermented a season of discontent, his secret organization, the

Arya Bandha Samaj which was located at Wardha, eighty kilometres south-west of Nagpur, becoming a fountain-head of inflammatory nationalistic propaganda. Clandestine emissaries travelled from Wardha to Poona to Bengal to Bihar. A 'Defence Fund', set up to secure Tilak's release, found supporters among leading Bengalis—men of letters like Rabindranath and aristocrats like Hem Chandra.[35]

When the Shivaji festival was held for the first time in Calcutta in 1902,[36] the *Times of India* Editor Valentine Chirol was provoked to say, 'In the Deccan the cult of Shivaji, as the epic hero of Mahratta history, was intelligible enough. But in Bengal his name had been for generations a bogey with which mothers hushed their babies, and the Mahratta Ditch in Calcutta still bears witness to the terror produced by the daring raids of Mahratta horsemen. To set Shivaji up in Bengal on the pedestal of nationalism in the face of such traditions was no slight feat...'[37]

Yet there was no reason for the British to feel unduly perturbed. As late as 1902, nationalist leader Bipin Chandra Pal was declaring '...we are loyal...because we believe that God himself has led the British to this country, to help it in working out its salvation, and realise its heaven-appointed destiny among the nations of the world. And so long as Britain remains at heart true and faithful to her sacred trust, her statesmen and politicians need fear no harm from the upheaval of national life in India...'[38] At the Town Hall meeting in Calcutta called to mourn the death of Victoria in 1901, prominent citizens had solemnly responded to the Viceroy's call.[39]

In the spring of 1904, handsome twenty-five year-old Subodh Chandra, nephew of nobleman Hem Chandra Mallik, having returned from a spell at Cambridge, was settling down to the life of an anglicized Bengali aristocrat at the family home at Wellington Square. Fabled for lavish hospitality, the Wellington Square mansion would often be ablaze with the lights of gala entertainment. Outside, the line of carriages would belong to both European and Indian dignitaries.[40]

But the passing of the Universities' Act came as a rude jolt. On 1 June 1904, Subodh and his friends established the Field and Academy Club on the ground floor of a house at 16 Cornwallis Street in north Calcutta. It soon became a platform for militant nationalists.[41] Some time later, when Lieutenant-Governor Sir Andrew Fraser decided to pay a visit to the Wellington Square mansion, ostensibly to see the Mallik art collection, Subodh Chandra refused to be present to receive his guest. Putting a younger cousin in charge, Subodh pointedly left the house.[42]

As unrest quickened across Bengal, columns of protesting students could be seen marching down Calcutta streets. News was brought of mass meetings held in Calcutta and the towns and districts of east Bengal.[43] But scornful of what he dismissed as hysteria, the Viceroy loftily told the Secretary of State, 'You need not be in the least disturbed at the popular native outcry against the Universities' Bill, it is largely manufactured.'[44]

Even the ever-faithful Lord Ronaldshay, Curzon's official biographer, was provoked to say, 'The fact of the matter is that Lord Curzon reserved to himself the right to decide when public opinion was an expression of views based on sober reasoning and supported by obvious justice, and when it was a mere frothy ebullition of irrational sentiment.'[45]

Chapter 14

'They Call Him Imperial George ...'

'They call him imperial George...,' lamented Sir Walter Lawrence, the Viceregal Secretary.[1] The two had been friends since their halcyon days at Oxford, Curzon branching forth into politics, Lawrence making a name for himself in the ICS. Topping the examination in 1877, Lawrence came out to India. But after a stint, when he returned home to work for the Duke of Bedford, Curzon coaxed him into accepting the job of Viceregal Secretary.

It was not an easy appointment. No sooner had the Curzons landed in Bombay than the difficulties started for Lawrence. Curzon was soon displaying 'signs of an awkward temper which showed his unfitness for the office he had taken up,' said Lawrence in dismay: the Viceroy-to-be had 'raved because the ADCs at Government House did not know that he had someone gazetted as Lt.-Col.'[2]

In Calcutta, Curzon had censured the authorities 'for the slovenly turn-out of the troops,' even before he had taken command. Lawrence wondered ruefully, 'I thought till the Viceroy was sworn in we had nothing to say to anything.'[3] As the Curzons settled down in Government House, problems multiplied. The staff was constantly bickering and there were frequent rows. But Lawrence loyally attributed the troubles to the new experience of officials with 'a Viceroy young, resolute and hardworking. He can beat them all at work and is always sure of his facts... they dislike his frank criticism and resent any criticism as interference.'[4]

However, a more damaging tale about the viceregal couple was soon doing the rounds. 'A common story served in England is that the consort has to curtsy in the morning.'[5] The origin of this story probably lay in Mary's excessive zeal for the ceremonial surrounding her husband's office. Calcutta society was embarrassed by the young American Vicereine's unabashed delight in the somewhat starchy protocol; Mary had happily agreed to the custom that dictated that she, 'jump up' every time her husband walked into a room, to keep standing until he first sat down and to be on constant guard 'to jump up every time he does'. The custom then was for 'the Viceroy and his wife to walk side by side', but the over enthusiastic Mary saw to it that when they marched into big crowded rooms, she went in 'several paces behind'. After all, as she confided to her father, 'George is... like a reigning Sovereign... the only difference is that he has a great deal more power than most kings...'[6]

In due course the stories travelled back to England providing grist for the public mill. They came to Mary's ears in the summer of 1901 when she was on vacation there, and she promptly sat down to complain to her husband about '... the silly tales about our foolish formality & stiffness in India.'[7] But he scoffed at them saying, 'it is all jealousy', reassuring her 'your beauty & my success have given rise to the wildest inventions'.[8] Instead of attempting to squash the rumours, the Viceroy '... courted publicity for his actions'.[9] Naturally 'stories clustered around him... He was a conspicuous target and seemed to invite arrows.'[10]

The Curzons seemed to have everything going for them— good looks, presence and the highest social position outside England. Freed from the confining strictures of Westminster,

Curzon had felt free to go his way, playing out the role of a British moghul, as it were, to the hilt. With unencumbered blithe spirits and an adoring and indulgent wife who also relished the trappings of an oriental court, the horizon was unlimited. But the twenty-eight year-old Mary had no experience or training in British public life. The dramatic ascent to being vicereine had given her little time to develop ease in the society in which she was overnight called to preside.

She was young, she was beautiful, she was elegantly attired, she could be vain. Despite two pregnancies she remained marvellously slender, her waist 'so small that 40 years later her own twin granddaughters could not hook the eyes of her lovely gowns'.[11] She had a penchant for jewels and expensive Paris clothes which she wore with the ease of a millionaire's daughter. Newspapers hounded her. 'I wore a white gown with white flowers embossed in velvet on it,' she said with an injured tone, when the papers had printed 'foolish accounts of imaginary bejewelled dresses covered with real stones'.[12]

She found the British women 'chilly' and 'critical' and told her mother, though not without relish, how '. . . all English women are ready to pounce as my being here excites such jealousy in many hearts'.[13] They, on their part, probably resented having to accept this exquisite young creature as their Vicereine. Many of them enjoyed old India connections and hailed from blue-blooded English families and so probably huffed at having to curtsy to an American girl.

Being an American did not help matters. Mary had been anxious '. . . to keep up the dignity of this position' and to show 'how nice and quiet Americans could be'.[14] But alas! Her natural exuberance came in the way and she was making her husband's starchy stiff-upper-lip colleagues squirm by her 'frankness of expression'. One official recalls an occasion when he was placed beside Mary at a lunch where a number of distinguished visitors were present:

> Their Excellencies were about to pay a visit to Burma. She turned to me and asked what was behind what the Rangoon papers were saying about a storm on the horizon 'no larger than a Burmese

lady's petticoat'. I explained that one of the high officials there had married a Burmese wife, and there was an awkward question about her presence at State functions. 'What do you think', she asked, 'of that sort of thing?...But why marry her?'[15]

To aggravate matters, Mary's family also descended upon India no sooner had the Curzons settled down in Government House. 'Socially the advent of the Leiters does great harm,' Lawrence groaned prophetically. Loud, vulgur, flirtatious, the Leiter sisters, Nan and Daisy, had got into boisterous and easy familiarity with the viceregal staff. Anticipating trouble, Mary had warned her father, 'I won't allow any flirtations as here I am a kind of Queen of Seringapatam and can't have flirtations at my court!'[16] But her exhortations fell on deaf ears. Nancy shocked the staid company at the Viceregal Lodge in Simla by pointedly asking the young ADC wearing very short breeches '...if he had grown out of his *pants*!' Daisy had raised eyebrows when she promptly 'annexed Captain Meade, quite a harmless youth on the staff'. Mary was embarrassed by the 'huge sham rows' of pearls they wore to parties, which she found 'very vulgar'.[17] In no time at all the girls had put on so much weight that even their four year-old niece Irene was calling them 'Aunt Elephies'.[18]

More horrifying was the practical joke the girls played on their brother-in-law. Tickled by the punctilious adherence to protocol at ceremonial functions, the naughty girls had solemnly marched up to the Viceroy and prostrated themselves in mock salutation at his feet.[19] That too in the presence of a vast gathering of people. Mary could have died of shame.

Nevertheless, George and Mary continued to 'do everything for their [sisters] enjoyment that mortals can.'[20] The viceroyalty, as the Curzons were discovering, was proving to be an expensive affair. They had been forced to incur an expenditure of £9,719 4s. 6d. in coming out to India. This included fares, freight, insurance, stores, wine, horses, carriages and harness. They then had to take over the stables from their predecessor Lord Elgin at the staggering cost of £5,500. To meet these charges they got £3,500 from the government and £500 from Mary's father. Mr Leiter later had to advance a further £2,000.[21] Enjoying their visit, Nan and Daisy stayed on with the Curzons for almost a year. The helpless Mary got 'an awful scathing in the papers and in society for "giving my sisters undue importance".'[22] But worse was to follow. Unknown to the Curzons, 'They abuse the V [Viceroy] behind his back and publish all the inner life of Viceregal Lodge,' Lawrence reported, adding 'I shall never forget the incident of... Miss D. I. [Daisy Irene] saying of... [presumably their sister Mary], "why, she is a greater liar than our mother" this in the presence of the staff. The situation was all through the year very painful and difficult.'[23]

Curzon's ADC, Raymond Marker, said of Mary's mother, 'I have become acquainted at last with Mrs Leiter—who is really a nice old lady and her "malapropisms" have been greatly exaggerated—I do not think she does more than average in the way of saying unfortunate things.'[24] But at that point of time, the dashing Raymond Marker was wildly infatuated with Daisy and had proposed marriage.

The match was later broken off, supposedly at Mary's intervention; it being believed that Mary had aimed higher for a sister of a Vicereine. Daisy eventually settled for the Earl of Suffolk, another of Curzon's ADCs. But the damage was done. Marker never forgave the Curzons, holding them responsible for his having been jilted. Marker was later at the War Office in London when it was at battle with Curzon; Marker did his utmost to take revenge.[25]

While these stories of the viceregal court were current, news had come of Sir Mackworth Young, the Lieutenant-Governor of the Punjab, having publicly quarrelled with the Viceroy over the taking away of North-West Frontier administration from the Punjab Government. Curzon may have had valid administrative reasons on his side but he had failed to take the Governor into confidence before passing the order. Hurt and angry, Young had publicly lashed out at the Viceroy at a Masonic Club banquet in Simla. Though the Viceroy was his superior in India, the two shared a common bond—they were both Eton men. When Mary's sisters had come out to India, it was Lady Young who had taken turns to escort them to parties and balls. Sir Mackworth's pride had been deeply wounded by the Viceroy's thoughtlessness, but once his anger had subsided he had apologized and tried to make amends by issuing an invitation. Had Curzon been able to nobly rise to the occasion and let bygones be bygones, he could have emerged from the controversy with his prestige shining. Unfortunately, he chose to act in a petty manner.

But even had Curzon wanted to forget and forgive, Mary was on the warpath. Incensed that the Youngs should dare to challenge her husband's authority, Mary upbraided him for not having been more peremptory with them. 'The L.G's remarks are so insolent,' she had cried in fever pitch, 'that I cannot understand your remaining on friendly terms with him; he slanders you in public & he shows you no respect whatever & I boil with indignation at him dining next night at Govt. House.'[26]

Curzon hastily assured Mary:

The Youngs had their first ball on Monday. Neither I nor Lawrence nor Baring went. You need not be afraid darling of my showing insufficient dignity in the matter. I accepted the apology because I had no other alternative, & because otherwise M.Y. would have had to resign, & then there would have been a hullaballoo in the Indian & English press. I should have emerged from it alright but it would have been a troublesome episode & I should never have been able to tell the truth of the case. Also it would have converted M.Y. whom everybody without exception has condemned into a martyr with whom everybody sympathised. It was despicable therefore to make it up outwardly. But I have never spoken to him or to her since, and I shall never enter their house again. Would you believe it that woman—whose continued abuse of me is the talk of Simla—had the effrontery to come up again a few days ago (after I had refused both her dances) with a renewed invitation to dinner? In reply I need hardly say the Viceroy regretted that he could not add to the number of his engagements. I should hope they see it now.[27]

In his need to pacify his wife, Curzon forgot that he was a

Viceroy of India dealing with his Lieutenant-Governor.

Young appealed again to the Viceroy to accept the invitation, even sending a copy of a letter Lady Young was supposed to have written to the Viceroy. 'I can't help noticing that throughout this season Your Excellency had entirely withdrawn the sunshine of your favour from me and naturally I am sad about it.'[28] Young pleaded, 'Any slight shown to her cannot be kept secret but becomes public property, detracts from the Lieutenant-Governor's social position and causes a deplorable scandal in a society where the Viceroy counts for so much.' He begged the Viceroy to resume social relations, 'as the King's representative and as a gentleman, in the name of justice and chivalry...'[29]

But freshly spurred by his wife's exhortations, Curzon refused the olive branch, provoking T. Raleigh, a member of the Viceroy's Council to say,

> I have no right to offer advice to H. E. except on this ground, that it will, in my opinion, be a public misfortune if Sir Mackworth, leaves Simla for the last time, after 40 years of service, without any friendly notice from the Viceroy. And may I perhaps be allowed to say that, *as far as my knowledge goes*, I don't think H. E. would suffer any loss of dignity by accepting an invitation at this moment. It would be understood in Simla that the reconciliation is more or less formal and the Viceroy's reason for agreeing to it is the public reason which I have indicated.[30]

It was several months later, when Young was retiring, that Curzon penned him a letter: 'As Head of Government of India, I should not like you to leave this country as you shortly will, without writing you a parting word of grateful recognition...'[31] But it was too late. The damage had been done. Wounded to the core, Sir Mackworth Young's sullen reply was: 'I must remind you that you have thought fit to treat me with marked discourtesy on a charge the evidence for which you will not disclose,...that I have appealed to you as the King's representative, and as a gentleman, to make the required amends, but without avail; and while matters stand thus between us, I am unable to accept with any satisfaction Your

Excellency's appreciation of my humble services.'[32]
Curzon's peevishness was never forgiven.

Shutting his ears to the stories, Curzon began to hector the bureaucracy on his ideal of what it should be. He said:

> The keys of India are not in England, nor in the House of Commons. They are in the office desk of every Civilian in this country. He by his character and conduct, is insensibly, but materially, contributing to the future maintenance or collapse, of the British dominion in India. If he is like the men who went before him, if he is keen about his work, has a high sense of duty, and is interested in, and likes the people, our position will be secure for a century to come. If he is indifferent, or incompetent, or slack, if he dislikes the country and the people, and has no taste for work then the great structure of which we are all so proud will one day break down.[33]

Following the Mutiny, the administrative machinery in India lost its pioneering zeal, and the civilians, though on the whole diligent and incorruptible, had felt the best way to govern India was to do the minimum mechanical task. Within months of his arrival in India, Curzon was despairing over the governmental machinery, comparing it to a 'gigantic quagmire or bog, into which every question that comes along either sinks or is sucked down; and unless you stick a peg with a label over the spot at which it disappeared, and from time to time go around and dig out the relics, you will never see anything of them again.'[34] Evan Maconochie was to say admiringly of the youthful Viceroy:

> One of his earliest orders was to the effect that every officer who noted on a case should sign his name in full. This transformed us at once from clerical shadows into actual beings, with bodies to be kicked when necessary and souls, as it might happen, to be saved. Nor was it long before the startling conception of an Under-Secretary's view being preferred to that of his chief was, once in a glorious moon, realised in fact.[35]

But old India hands did not like the Viceroy's exacting style and hoped he would soon tire. When a Calcutta merchant told a Government of India secretary that 'this new Viceroy will hustle you secretaries,' the confident rejoinder had been, 'No, he will be paper-logged in three months'. But the civilian was proved wrong. It was the secretaries who were paper-logged, while the Viceroy sat at his unencumbered table asking for more.[36]

But having taken on the task of overhauling the administrative machinery, Curzon could not bring himself to parcel out the work. He scolded his subordinates for apathy and took on all the work-load on to his own shoulders. When Hamilton, Secretary of State for India, cautioned that he must not unduly tire himself performing labours which could well be done by subordinates Curzon grumbled:

> I must thank you for the constant and friendly insistence with which you press me not to overwork. But when you urge me to delegate more to my subordinates, I confess I think you hardly realise how quickly such a proceeding on my part would bring the whole machine on a standstill... I tell you it is perfectly useless. The Government of India is a mighty and miraculous machine for doing nothing.[37]

Infuriated senior civil servants, interpreting the Viceroy's refusal to delegate work as an insult to their competence, felt that the superior Viceroy was so conscious of the inferiority of others that he had to do everything himself. Tongues wagged more furiously. Reports reaching London said there was 'much whimpering from Army, & disgruntled civilians.'[38] As Lawrence observed, official India 'rather resented the young man in a hurry, who had discovered that all was wrong before he came; they also resented the quickening of the pace.'[39]

The refusal to delegate responsibility widened further the chasm between the Viceroy and his staff, which drove Curzon to work even harder and get even more irritable. Thus the vicious circle grew. St. John Brodrick warned Curzon early in his vieroyalty to guard against such resentment. His warning was friendly and full of solicitude for Curzon:

> Once or twice you have alluded to the unpopularity your reforms are causing you in certain quarters, and the lies as to your social proceedings which find their way into the English Press. You are now still only one-third through your time, you have big schemes

on hand, and I can't bear the thought that you are in any avoidable way handicapped...Briefly—I gather you are raising against yourself an undue amount of opposition, on grounds more personal than political. You drive some hard and they resent it. You rebuke others and they cannot turn. You ignore a few and their mulishness reacts on others. You and Mary started *in excelsis*. We have got a young god and goddess come to rule us. There was bound to be a reaction. But surely, making every allowance for misquotation of your words and mis-statements as to your actions, you ought, with your great hospitality and personal charms, to enjoy more popularity than you do.

You cannot run so vast a business as India off your own bat. Physical strength can do only a certain amount. I can't bear to think of you for three years, with health needing care, pulling double against a sulky staff. What do I suggest? Merely this. Everyone respects your powers of work, your insight, your decision, your courage. They fear you, I am told. You have therefore an asset on which to play. Why not smooth everything for a bit? People will be the more grateful, if they are a little sore just now.[40]

But turning a deaf ear, the Viceroy slogged fourteen hours a day. He wrote his own speeches, he drafted his own bills. A colleague said despairingly, 'He would revise his chef's menu for dinner parties, and fix tunes to be played by his orchestra. I have seen him, bedridden and evidently in much pain, carefully revising the phraseology of a draft letter to a provincial government, which would be read by no one of higher rank than an under-secretary.'[41]

Overwork and the excruciating pain in his back brought him nearly to the end of his tether. When Mary was away in England in the summer of 1901, Curzon wrote to her in tones of great self-pity, 'Poor Pappy [her pet name for Curzon] gets so downcast sometimes...grind, grind, grind with never a word of encouragement...I am crying now so that I can scarcely see the page.'[42] But he had brought it on himself.

As Viceroy he even took time off to supervise his family expenditure. Mary told her parents, 'Even here George helps & sees to all accounts and works like a slave; it is no use begging him to work less as he will work till he kills himself in spite of all the begging not to in the world.'[43]

But the energetic Viceroy could not resist the temptation to check the spoons and sack the servants. There is the extraordinary story of how on one occasion the Viceroy personally set himself out to catch a kitchen thief. Disturbed by what once seemed to be an abnormally large requisition of chickens by the Government House kitchen, the Viceroy, like a detective, made systematic inquiries. From the suppliers he found out that the Exchequer had been called upon to pay for 598 chickens when the requirement was only for 290. Having discovered the fraud, the Viceroy put himself out to personally nab the culprit. Triumphantly, he later reported to a friend in England, 'We caught him red-handed'.[44] What is perhaps even more astonishing is the shade of vengeful triumph in the Viceroy's voice. The villain, had been spotted, cornered and finally caught; and the bourgeois self-righteous Victorian temperament had found an outlet. In this need to ferret out, probe, chastise, Curzon was no different from others of his age, though he tended to be disappointed when such an opportunities did not come his way.

A senior official wrote of one occasion when Curzon had asked him to share lunch. Overburdened with his work, the official excused himself. But the Viceroy was much offended, turned his face away, and would not look at the unfortunate official for three or four weeks. The officer was shocked that the Viceroy's 'greatness could exist along with a certain littleness of mind'.[45]

Pettiness, an inability to delegate to others, a fussy preoccupation with detail, were then generally believed to be attitudes in marked variance with those associated with people belonging to Curzon's class. Margot Asquith had branded them as part of Curzon's 'middle class method'.[46] Curzon's father, Lord Scarsdale, having unexpectedly come into his inheritance, had never been able to live with the easy abandon of an aristocrat. Some of his insecurity must have rubbed off on his son. Curzon's daughter Irene wrote in her autobiography, 'My father engaged his first valet when he was twenty-seven years-old, and was worried at the wages he might have to pay'.[47]

Curzon's grandfather was only the seventh son of the 2nd Baron Lord Scarsdale and, had it not been for the illegitimacy of his five older brothers, Curzon might have remained just such an ordinary country gentleman. That is why Curzon always believed that it was his destiny which had brought him into his inheritance. He was caught in the manner and prejudices of the day. Though he tried to maintain a carefully calculated patrician style, beneath the façade he was but a busy bourgeois.

Perhaps Curzon's colleagues in India might have understood him better and been more inspired by him had he tried to grow closer to them by occasionally sharing his fears and hopes. Few, even then, for instance, knew that Curzon's stiff gait, which lent such an awful majesty to his presence, was the result of that leather and steel harness worn since his childhood mishap. As he wrote to his father:

> The fact is my back is a handicap to me such as no one can possibly realise. I always have to wear these vile stays which make me stream with perspiration when everybody else is quite cool. I am constantly in discomfort and sometimes in pain when I have to appear entirely at ease. No one who saw my back ... would think it possible for me to do one tenth of what I do. I do it now, because I never can feel sure that in 10 or 15 years time after I have returned to England—or earlier—I shall be good for anything. I never breathe this to a soul.[48]

Had he disclosed this to his subordinates they might have tolerated his bouts of rage as a human failing.

As it was he was lonely in his eminence, isolated as he was by his position, if not by his own nature, and had no one except his wife to whom he could confide his fears, anxieties and hopes. To all others he was the Viceroy and an exacting master. This knowledge frightened people. Instead of trying to make them feel augmented by his presence, he made them feel less than what they were. Unlike with Winston Churchill, few were allowed to see the man behind the mask:

> The contradictory qualities which dwell in the characters of so many individuals can rarely have formed more vivid contrasts than in George Curzon. The world thought him pompous in manner

and in mind. But this widespread and deep impression...was immediately destroyed by the Curzon one met in a small circle of intimate friends and equals...Here one saw the charming, gay companion adorning every subject that he touched with his agile wit, ever ready to laugh at himself, ever capable of conveying sympathy and understanding. It seemed incredible that his warm heart and jolly boyish nature should be so effectually concealed from the vast majority of those he met and with whom he worked.[49]

Few knew of the pain that racked him, robbing him of sleep at night and turning him into a cantankerous chief by day. Unaware of his ordeal, they ascribed his impatience to a viceregal disdain for lesser mortals. 'In public he was the great, stately Proconsul, sonorous and formal. In his own house he would often throw off the mantle and be natural...He would say things in jest—often taken in earnest by those who watched the handsome stony face and were ignorant of his love of laughter and jokes,' said Lawrence.[50] He could have and ought to have at least attempted to hear out his closest subordinates. In this lay his chance for survival, both as an individual and as a Viceroy. But pathetically anxious to retain his image as a 'superior person', Curzon did nothing to refute the myth. On the contrary, he drove himself to retain an aura of distant eminence. He projected to all those he worked with the image of a lofty and arrogant man: he felt it was expected of him and he pushed himself to perform what was expected. Friend and diplomat Viscount D'Abernon said, 'The Marquess Curzon was born grandiloquent,' adding that he could give 'orders to a footman in a language which would not have disgraced Cicero addressing the Roman Senate'.[51] Curzon certainly had that archaic grand style, but it was a cultivated style. Curzon's deputy at the Foreign office, Lord Vansittart, was perhaps more perceptive when he noted that it was not a natural gift, but that Curzon had deliberately 'designed himself as the most anachronistic of "grand signeurs" outlined against the sunset of the breed.'[52]

It was by clinging to a value system valued in an earlier era of stricter class demarcations that Curzon found it possible to survive. Taking on an overwhelming work-load he brought himself to a breaking point. But he could look around at his Indian Empire and feel at last he was valiantly fulfilling the role in life for which he convinced himself he had been predestined.

Chapter 15
A Fateful Appointment

In bringing to India the tough and ruthless Kitchener as his Commander-in-Chief in the autumn of 1902, Curzon seemed to have deliberately courted danger. Almost systematically Curzon antagonised those around him—civilians, the Army and the Indian social and political élite. Kitchener's arrival set Curzon on a collision course with the Home Government which was to lead to his final departure from India. Well meaning friends had warned Curzon of the dangers, Rennell Rodd among them:

> Kitchener has many faults—he is not straight in the sense that having had everyone's hand against him for so long, he tries to achieve his objects in round about ways, mistrusting the direct and open method; he is secretive and not frank; . . . he is even, if one must say it, unscrupulous in his methods.[1]

The Secretary of State wrote, 'After some communications with my colleagues, your opinion carried the day and it was determined to select Kitchener for C-in-C in India. I look with some apprehension upon this appointment, as I fear the effect of his rough and unsympathetic manner and strong economic hand upon the native army. You will have to watch him.'[2]

But as always, the greater the challenge, the greater was Curzon's eagerness to meet it. Kitchener of Khartoum had become a national hero after he had dramatically avenged the death of General Gordon, destroying the Mahdi, hacking down the dervishes and finally turning the tide in favour of the British in South Africa.

Kitchener's reputation as a bloodthirsty operator was no secret. The press had made much of his barbaric treatment of the Madhi's body. Four days after the Battle of Omdurman, Kitchener issued orders to ransack the Mahdi's tomb and carried away the skull as a personal trophy.[3] For a while he had toyed with the idea of mounting it on silver or gold and deliberated over whether it could best be used as an inkstand or as a drinking cup, until Lord Salisbury cabled him, 'The Queen is shocked by the treatment the Mahdi's body has received, and thinks the head ought to be buried'.[4]

As usual Curzon had chosen to ignore the warnings, confident that the statesman could any day handle a soldier. Besides, he was also conscious that in bringing Kitchener to India, he was elevating his own position, for the great imperial hero would be subordinate to the Viceroy on the Indian soil.

Unlike Curzon, Kitchener had not been born into the nobility. Nine years older than Curzon, Kitchener was an arrogant, rough, immensely tall and powerfully-built man who seemed to tower over his contemporaries. From his great height he gazed down at the world with a formidable aloofness. The bleak narrow eyes were set in a square face partly masked by his proudest possession, his moustache. In part, this imposing bush also helped to compensate for the coarseness of his skin which had made a colleague compare him to 'a rough pirate'.[5]

Kitchener was the son of a colonel who sold his commission in the army to buy land in Ireland which was going abegging due to the great potato famine of 1846–7. Kitchener's biographer Philip Magnus said of the father, 'He was a disappointed and frustrated man, noted for the foulness of his language; and he became an eccentric martinet who ran his home as far as possible on army lines. His hour for breakfast was eight o'clock, and no servant would have dared to be even a few seconds late.'[6] One of his eccentricities was to pave the way for Mrs Kitchener's early death. Averse to the use of blankets in bed, Colonel Kitchener insisted on substituting them with sheets of newspaper which he felt were better conductors of heat. 'Accordingly, around the double bed in which he and Mrs Kitchener slept, boards were erected to which sheets of newspaper, sewn together, were attached.'[7] Mrs Kitchener was to catch a fatal chill.

Colonel Kitchener also detested schools and his children were trained privately by tutors. In 1868 Kitchener passed out of the Royal Military Academy at Woolwich. He was eighteen years old but had no formal training, and did not know the meaning of team spirit. In childhood he had been subjected to rigorous punishments. He submitted himself to being spread-eagled on the lawn in the attitude of a crucified man—'on my back in the hot sun, pinioned down by the arms and legs which were roped securely to croquet hoops,' recalled Kitchener.[8]

Choosing the Army as a career, Kitchener made rapid advances. His military acumen and personal courage in battle were unmatchable. His notable success in South Africa about the same time as Curzon's appointment to the Indian viceroyalty had raised Kitchener to the level of one of the greatest British soldiers. Victoria admitted him to the peerage as Baron Kitchener of Khartoum. This opened the doors of Britain's great country houses to him. One such door was that of Salisbury's daughter-in-law, Lady Cranborne.[9] Finding in her an ardent admirer, Kitchener had set himself to cultivate her. In his years in India, Lady Cranborne proved to be his valuable link with the Home Government.

A man of rough and ready methods, Kitchener ordered the rebuilding of Khartoum by ordering his officer to 'loot like blazes. I want any quantity of marble stairs, marble paving, iron railings, looking-glasses and fittings; doors, windows, furniture of all sorts.'[10] Similarly, when he finally left South Africa he was laden with a great quantity of loot. He carried away life-size statues of Kruger and other prominent Boers which had been removed from public squares in Bloemfontein and Pretoria under his personal orders. He meant to decorate his private park with them when he acquired a house in England.

The two protagonists had much in common. Both could be absorbed by fussy, domestic detail. Both were arrogant, high-handed men not capable of playing second fiddle. Both relished stately living. Unlike Curzon, however, Kitchener had tact and cunning and was adept at building up a chain of contacts who would help him to further his cause. In England he had charmed not merely the Prime Minister's son and daughter-in-law, but had also made friends with Balfour and assiduously cultivated the Duke of York who later became King George V. In India too, he was to rapidly build up a series of confidants. Curzon, by inflicting punishment on the 9th Lancers, had dug a deep chasm between the Viceroy and the Army. The situation was ripe for Kitchener and he was not slow to seize it. Though Kitchener never married, preferring to surround himself with a band of young unwed officers, he was capable of exercising great charm with the ladies and even Queen Victoria relished his company.[11] In fact, even before Curzon had sailed to India to take up his viceroyalty, Kitchener had indicated his desire to join him there. Kitchener had concluded his letter saying, 'I enclose a photo for Lady Curzon, to remind her of the man who means to take her down to dinner some day in India'.[12]

In 1900 Curzon wrote to Hamilton, the Secretary of State, saying that in India, 'We want a Kitchener to pull things together'.[13] He envisaged grave threats to India's integrity from Russia, China, Afghanistan and France, and had despaired of the Home Government's apathetic attitude in tackling it. 'I do not suppose that Lord Salisbury will be persuaded to lift a little finger to save Persia from her doom. . .,' he had grumbled.[14]

Kitchener arrived in December 1902 and joined the Curzons at Bhuratpore. Curzon reported to the Secretary of State that his new Commander-in-Chief 'greatly impressed me by his honesty, directness, frank common sense, and combination of energy with power. I feel that at last I shall have a Commander-in-Chief worthy of the name and position.'[15] The very next day, however, Curzon was saying that Kitchener told him 'he had made a a mistake in coming out as Commander-in-Chief, and he thought he ought rather to have been Military Member. In his view the Commander-in-Chief ought to be the Chief Military Adviser to the Viceroy, instead of which it seemed that the position belonged to an officer of inferior experience and rank.'[16] The Viceroy said he asked his Commander-in-Chief to wait a little and see the system in practice.

In India, as in Britain, the system had been one of dual control in the army, with the Commander-in-Chief being in charge of manœuvres, distribution of troops, promotions, intelligence and discipline; matters concerning finance, stores and supply were vested with the Military Member. Though inferior in rank to the Commander-in-chief, the Military Member had a seat on the Viceroy's Council and powers to veto the Commander-in-Chief's recommendations. The Viceroy was unaware that Kitchener had made a quick survey of the situation and was moving towards doing away with the position of the Military Member.

On 25 January 1903, Kitchener was complaining to Lady Cranborne about this 'extraordinary' system: 'I asked Curzon why he liked to keep up this farce, and his answer was, "If the Commander-in-Chief had anything to do with the machinery, he would become too powerful".'[17] By February 1903 Kitchener had showed his claws. He refused to sign a document drafted by the Military Member, Sir Edmund Elles, and backed his refusal with a threat to resign.[18]

Hitherto, neither of his military men had posed any threat to the Viceroy. The earlier Commander-in-Chief, Sir William Lockhart, suffering from a long-drawn illness, had collapsed and died a few hours before returning to England for treatment in the spring of 1900. About his Military Member, Sir Edward Collen, the Viceroy's comment had been, he 'had dangled his sword for the best part of a lifetime from an office stool'.[19] The two men had sparred between themselves; little intrigues and battles which the Viceroy had watched with amused indulgence. Reporting to the Secretary of State in 1899 about them, Curzon had said:

> I hear that a novel and local storm is slowly brewing in the 'arcana' of the military bureaux themselves. The Commander-in-Chief is said to be evolving a scheme for the abolition of the Military Department; and meanwhile I hear that the Military Member, all unconscious of his impending doom, is elaborating a counter-scheme for the extinction of the Commander-in-Chief. It looks as if I, who am a consistent though amicable antagonist of both, would ultimately have to step in to save them from mutual destruction at each other's hands.[20]

With something of a rude shock Curzon now became aware that in Kitchener he had an entirely different kettle of fish. Curzon's second thoughts about the nature of his Commander-in-Chief came too late. Kitchener sent secret letters through his staff to Lord Roberts, the War Secretary, and to George Hamilton. Curzon was in Simla when he received a letter from the Secretary of State in which he said he had heard form Lord Roberts at a meeting of the Imperial Defence Committee in London that a 'scheme of wide reform' was put by Kitchener.[21] Curzon accosted Kitchener:

> In his letter which came by yesterday's mail, the S. of S. [Secretary of State] mentioned that he had heard from Members of the Imperial Defence Committee of 'Schemes of Wide Reform and of great alteration' being put forward by you—as he assumed in private letters to the W. O. [War Office] or C-in-C; and he asked me to warn you that—although communications between the two C-in-Cs are always recognised, any changes of an important character must be referred through the Indian Government and the India Office here. Otherwise we shall have a double set of communication which will be a source of great embarrassment and personal friction.[22]

Kitchener accepted the rebuke but began intensifying his campaign with the Prime Minister through Lady Cranborne, begging her to be very careful and consider everything he wrote as private.[23] Curzon still refused to seriously register the danger signals.

Outwardly a truce was declared. But both protagonists knew that if Kitchener threatened resignation, he would not win because of any weakness in Curzon's case or strength in Kitchener's but because of the latter's great public prestige. Curzon confessed so in a letter to the Secretary of State, saying that if Kitchener were to resign, 'Public opinion in England though certainly not in India would side with him'.[24] Nevertheless, Curzon still deluded himself into believing that he could control the soldier and referred to him as 'a caged lion, dashing its bruised and lacerated head against the bars'.[25] On 9 July Curzon was in Simla serenely reporting to the Secretary of State, 'the atmosphere as regards Kitchener has completely changed. He is out here with us in camp at this moment and not a cloud flecks the sky... He now realises his mistake and is aware that I am his best friend.'[26] What a fool's paradise the Viceroy was living in.

Curzon later wrote, 'During the remainder of the year 1903, and indeed until I left India for England in May 1904, the question of Military Administration in India was not revived.'[27] The truce, however, had come to an end with the announcement of the extension of Curzon's viceroyalty, Kitchener complained to his lady friend:

> The fact is no C-in-C worth his salt could go on with the military department organised as it is now... C-in-Cs can be provided, I have no doubt, for the pay, who will shut their eyes and let things go on; but I cannot; and as the Viceroy likes present system, there is no doubt I ought to clear out.[28]

Kitchener had little interest in leaving, having renovated his two residences in Calcutta and Simla so that they rivalled the magnificence of the viceregal abodes. At Treasury Gate, his official residence in Calcutta, Kitchener had brought down the inner walls to create gigantic halls for stately entertaining. Dinner was served at separate round tables. Here he entertained on a lavish scale, the showpiece being the massive gold plate which he had brought out with him. Besides five vases, there was a complete dinner service for six in solid gold.

A regiment of the finest cooks served his fare. A woman guest recalled, 'He must have a regiment of "cordons bleus" in the kitchen, for I never saw such a repast... We began with iced soup, just stiff enough to spoon comfortably, with little dots of truffle; next fillets of fish with mushrooms and prawns; then fillets de boeuf à la banquetine.' A 'mousse de canetons', followed by quails, constituted the fourth and fifth courses. The sixth was 'a dream of a fruit compote, with cream ices.' The seventh course was biscuits and butter and the eighth, 'such peaches, apricots, mangoes, and prunes, just softened with a dash of brandy'.[29]

In Simla, Kitchener had ordered his ADCs to pound up vast quantities of files belonging to the military department to be turned into papier mâché for renovating the ceilings in the library and the drawing room. He had the rooms panelled in beautiful walnut. Not content with the shine, he applied varnish to it. When the Finance Department protested at the expense, Kitchener browbeat them into paying the entire cost. Curzon could not have enjoyed the spectacle of his Commander-in-Chief trying to steal his thunder, but having brought it upon himself, he had no choice but to bear it. Simla society relished the tension between the two giants, and each little story was expanded and exaggerated before it was sent rapidly round official India and to friends in England.

Curzon's good friend, Sir Schomberg MacDonnell, scenting the danger signals, had said, 'I am however alarmed at your staying on in India; I feel it will wreck your health and render unfulfilled my dream of you as Prime Minister.'[30] He advised prompt return to England saying, 'Don't be ridiculous and pooh-pooh viceregally my counsel. I have so long been "Head-lad" of the Prime Minister's string that, I know the stayers among the field. And my dear George you are the only stayer among the lot... You have made your reputation as a great Viceroy and the time has come for you to direct other Viceroys.'[31] It was good advice and Curzon should have heeded it.

But Curzon was too busy with his imperial progress. The immediate project was to establish ascendance in the Persian

Gulf. He had bullied the Home Government into sanctioning a viceregal tour through the Gulf which was being now referred to as Curzon's Lake.[32] At Karachi he had stepped aboard the *Hardinge* with its escort of four naval ships, '... we sailed out of the [Karachi] harbour, to the booming of guns in the forts and manning of ships,' Mary said happily.[33] At Muscat the gun salutes were so deafening that they 'crashed and echoed against the harbour walls'. The town had been decorated with great festoons of flags; the Viceroy had been received by Sir A. Hardinge, the English Minister in Tehran, and the Sultan of Muscat. At Kuwait, the Sultan sent a special victoria driven by two Arab horses and imported from Bombay for the purpose. The Viceroy and his host set off in it for the tour escorted by a camel corps and about 250 horsemen. 'The Sheik's flag, with the timely inscription "Trust in God" sewn in white on scarlet background, was carried in front,' recalled Curzon.[34] As the procession had moved, 'it apparently became necessary for the cavalry escort to express their rejoicing not merely by war-cries of the most blood-curdling description, but by firing ball cartridges into the air.' In the confusion the British Minister was 'shot clear over the head of his steed and deposited with no small violence on the ground.'[35] But Curzon did not mind. He was the centre of a game of oriental ritual which this gave him enormous pleasure. At the sheikh's residence 'all the sons of the house were produced... and they ranged from 7 years of age to 50!' While there, Curzon also heard 'a sound of violent rending and tearing... Not a word was said on the subject' but when the Viceroy got back to the street it was to find the Bombay victoria 'reduced to matchwood... and the steeds vanished'. Apparently, the animals unused to being harnessed to a vehicle 'had made up for their orderly behaviour... by kicking the somewhat flimsy construction to pieces'.[36] Nevertheless, the viceregal party returned, 'quite enchanted with their adventures'.[37] In the midst of such attentions, MacDonnell's advice faded into the background.

In January 1904, Mary returned to England to await the birth of their third child. In April, Curzon had joined her there for a six-month sojourn before resuming the second term of his viceroyalty. He had convinced himself that there was a great task to finish in India—a task which only he could do. He told his brother that news of his second term 'has been very well received throughout India...and that I have done my duty despite of everybody'.[38] He had little inkling of the long knives being sharpened around him. Kitchener had intensified his lobbying in England. Balfour and Brodrick were not inclined to view Curzon's viceregal progress with the same indulgence. Brodrick might once have been his closest friend and have assured him of how happy he was to see his friend's success, but he was now jealous of the Viceroy who had leap-frogged him in the race for promotion. While Curzon was away from India Brodrick had urgently written to Kitchener, 'we have got to the point where it is absolutely essential that you and the Home Government should understand each other,' adding 'while the cat's away the mice will play'.[39]

In Calcutta, Kitchener ruled the roost. He frightened citizens out of their wits by thundering through the town, on the wrong side of the road, in a mail-phaeton drawn by a pair of magnificent black horses. Kitchener would hold the reins with only one hand, which also held his long cigar and a glove. The sentries on the bridge to Fort William fled for their lives at the approach of their Commander-in-Chief.[40]

Chapter 16
The Partition of Bengal

Curzon has not yet been forgiven in India for the partition of Bengal. Though he made this out to be an administrative measure to relieve an overburdened state, Indian opinion has always taken the Viceroy's act as a machiavellian measure to 'divide and rule' India by tearing apart the province along communal lines, giving birth to a Muslim majority province in the east.

The partition opened a pandora's box, letting loose subterranean forces in the nationalist movement. Overnight, a vast number of hitherto neutral people became politicized. Muslim opinion moved sharply away from the national mainstream, creating a platform for itself in the Muslim League. Nationalism and separatism began to strike hammer blows at the roots of British power and Indian unity, neither of which were consequences Curzon intended.

Spread over the lush green deltas of the Ganga and the Brahmaputra, Bengal has always been the golden or Sonar Bangla. From their capital Bikrampur, the Hindu Sen kings ruled from the eleventh to the thirteen century. With the Muslim conquest in the thirteenth century, the capital of Bengal shifted from Patna in the west to Dacca in the east. The coming of the British struck a blow to Muslim ascendancy. Regarding them as traditional foes, the British had deliberately discriminated against Muslims. Quick to take advantage of the new educational opportunities, the Bengali Hindu raced ahead of his Muslim counterpart, the rise of Calcutta as the capital of the Empire further enhancing the status of the Hindu *bhadralok*.

But now that Macaulay's[1] dream of an English-educated Bengali middle class who would be 'interpreters between us and the millions whom we govern—a class of persons Indian in colour and blood, but English in tastes, in opinions, in morals, and in intellect' was nearing realization, the authorities began to have second thoughts. In 1854 Charles Wood, the Secretary of State, said 'I do not care about young Bengalees reading Bacon and Shakespeare'.[2] Anti-Bengali prejudice among the British gathered momentum with Surendranath Banerjea's founding of the Indian Association as a representative body of the educated middle classes, a precursor of the Congress, with which it was later to merge.

The idea of partitioning Bengal was not new. For more than quarter of a century before Curzon's arrival, there prevailed a belief in official circles that Bengal was too large to be under a single administration and that its partition would lead to greater efficiency. In 1874 the province of Assam was stripped from Bengal and put under a Chief Commissioner. Though the new province carried away three Bengali-speaking districts—Sylhet, Cachar and Goalpara—the earlier partition did not arouse adverse comment. The idea of transferring Chittagong Division, in order to give a port to the eastern province of Assam, constantly cropped up, though it was never implemented. One Commissioner of Chittagong Division, W. B. Oldham, went further and recommended a larger transfer. Fearing the rise of the Hindu *bhadralok* as:

> ...politically threatening in the same way, if not in the same degree, as what is called the Bania Raj, which has been allowed to grow up in the Punjab and in Hindustan, and to be as anomalous as the claim made by the Poona Brahmins and Madras Mudaliars and Bengalis at large to competence and fitness for governing the martial races.[3]

When the scheme was suggested to Lord Elgin, characteristically, he sat on it, preferring to let sleeping dogs lie.[4]

The idea of partitioning Bengal was thus not Curzon's brainchild. In a letter to Hamilton in April 1902, Curzon had mentioned, 'Bengal is unquestionably too large a charge for any single man' and had asked whether Chittagong and Orissa should be separated. He favoured the creation of a Lieutenant-Governorship for Central Provinces swollen by the addition of Berar, Orissa and the Ganjam districts of Madras.[5] To his surprise he had found that Secretaries in the Viceroy's Council had been working on the subject for over a year:

> ... calmly carving about and rearranging provinces on paper, colouring and recolouring the map of India according to geographical, historical, political and linguistic considerations—in the manner that appealed most to their fancy ... Round and round, like the diurnal revolution of the earth went the file, stately, solemn, sure, and slow; and now, in our season, it has completed its orbit, and I am invited to register the concluding stage.[6]

Bureaucrats like Police Commissioner Andrew Fraser[7] and Herbert Hope Risley[8] had been busy working out a scheme to split Bengal. Perhaps they saw in Curzon the Viceroy best inclined to pick up so challenging a scheme. Shrewd civilians, they reckoned that Curzon would not be able to resist the temptation to plunge headlong where his predecessor had been reluctant to tread. The very fact that Elgin had not felt equal to the task, they calculated, was enough incentive to whet the Curzonian appetite.

There were other considerations. By dividing Bengal on communal lines and encouraging Muslim ascendancy in the east, the Government was, in Risley's words, weakening 'a solid body of opponents' to the Raj. The Viceroy had made no secret about his contempt for the Congress, having had no qualms about spurning the offer of Congress President Romesh Chandra Dutt to solicit the co-operation of Indians.[9] Curzon told Hamilton, ' ... the whole of our case against the party is this, that it is in no sense a representative national body, as it claims to be, and not actually disloyal to the British

Government in this country, it is at any rate, far from friendly towards it.'[10] He was determined to crush the power of Congress, not to do so would be to relax the vigour of the Government he had vowed to uphold. Declaring that the Congress was 'tottering to its fall', he said one of his ambitions in India was 'to assist it to a peaceful demise'.[11]

And here were the Secretaries offering the Viceroy, on a platter as it were, a scheme which would bring the Congress to its demise and that too under the garb of administrative expediency. What more could he want? Curzon happily fell in with their plan for dismembering Bengal saying:

> Give all thou canst—High Heaven rejects the lore
> Of nicely calculated loss or more.[12]

In 1905 Curzon was to tell Fraser, 'In the face of the chief offender, you seem to have got off very lightly about partition! But had the whole correspondence been published, which the Secretary of State actually proposed to do—I expect that you would have been crucified at my side.'[13]

Indeed, Andrew Fraser had been one of the masterminds of the operation. A member of the ICS, he had arrived in India in 1871 and was posted to the Central Provinces. In 1901 he was promoted to Presidentship of the Police Commission and in that capacity visited Dacca and Mymensingh districts of Bengal. On his return, Fraser reported how he 'was struck by their exceedingly defective administration and with the necessity of bringing them into contact with a strong personal government'.[14] Propagating the theory that these areas were 'hotbeds of purely Bengali movement which was seditious in character', Fraser began instigating Curzon towards their amputation from Bengal. His argument ran, 'I believe Dacca and Mymensingh would give far less trouble it they were under Assam, and also believe that East Bengal would not be so painfully prominent a factor in Bengal administration if this transfer were made.'[15]

The Viceroy in his Minute of 1 June 1903 acknowledged his debt to Fraser, whom he knighted and appointed Lieutenant-

Governor of Bengal. 'There remains an argument to which the incoming Lieutenant-Governor of Bengal, Sir A. Fraser attaches the utmost weight, and which cannot be absent from our consideration. He has represented to me that the advantages of severing these eastern districts of Bengal...'[16]

The Minute had proposed the incorporation of Berar into the Central Provinces, the retention of Sind with Bombay, the transfer of Chota Nagpur from Bengal to the Central Provinces, the union of the Oriya-speaking tracts and its retention with Bengal. Not only was Chittagong Division to be transferred to the new province of Assam and East Bengal but so also were Dacca and Mymensingh districts of the Dacca Division. While conceding that 'the change will be represented as one of a retrogade and pernicious character, tending to place a highly advanced and civilized people under a relatively backward and incompetent administration', the Viceroy glibly expected the opposition would 'rapidly dwindle... and before long disappear'.[17]

There was wide-spread protest throughout the country that the partition was politically motivated, public opinion not having missed the contradiction of Bengal being divided while Oriya-speaking districts were united and retained under a truncated Bengal.

The nationalist newspaper *Sanjivani* wailed, 'O Lord Curzon: O Sir Andrew Fraser: Pray... do not drive them [the Bengali people] from the bright and radiant land of Bengal into the dark and dire cave of Assam.'[18]

Appearing to cajole and appease opposition, Sir Andrew Fraser convened a conference over Christmas of 1903 at the Lieutenant-Governor's Italian renaissance palace, Belvedere, in Calcutta. In its elegant white and gold hall, sixteen landowners, Hindu and Muslim, met to confer. A Bengali leader said, 'I was under the impression... that the Government, as a result of these conferences, would bow to public opinion and withdraw from an untenable position. But this was not to be.'[19]

The largest Muslim landowner, the Nawab of Dacca, alert to the opportunities unfolding before him, moved to grasp them.

Photographs show a portly, thick-set man with a shining round face and dimpled cheeks. The Nawab was to advocate the inclusion of all the dominant Muslim districts of eastern and northern Bengal into the new province.

In February 1904 the Viceroy visited East Bengal, 'ostensibly with the object of ascertaining public opinion, but really to overcome it,' Banerjea said bitterly. Curzon was presented with hundreds of petitions which he chose to ignore. At Mymensingh, which was supposed to be transferred to East Bengal in the new scheme, the Viceroy was a guest of Maharaja Surya Kanti Acharya. While receiving the Viceroy with princely hospitality, the host minced no words about his displeasure over the partition.[20]

But in Dacca, the Nawab had laid out a red carpet. In his desire to woo Muslim support the Viceroy talked enthusiastically of how the new province with Dacca as its capital 'would invest the Mohamedans of Eastern Bengal with a unity which they have not enjoyed since the days of the old Mussalman Viceroys and Kings...'[21]

Yet Curzon could not be said to have previously been a special friend of the Muslims. Four years earlier, when they had protested at the Viceroy's insistence on the increasing use of Hindi for official purposes, Curzon had categorically ordered his officials to ride roughshod over Muslim sentiments. He had then said, 'The troubles of the Mussalmans merely represent the spleen of a minority from whose hands are slipping away the reins of power, and who clutch at any method of arbitrarily retaining them.'[22] But now imperial interest drew him close to the Muslims. Seeing that the Muslims were forthcoming in their support, Curzon moved rapidly in a bid to accommodate them. As a special concession to the Nawab, the Council unbent granting a Lieutenant-Governorship with a Legislative Council to the new province with its capital at Dacca.[23] As a further inducement, a loan of £100,000 was offered to the Nawab.[24] There had been little difficulty in gathering a vast crowd of Muslims to set the seal of approval on the Viceroy's plan. A British civil servant said of the Nawab that he,

... could not be described as a personage of special influence in his community before he took an active part in advocating this [partition] measure. He is not a member of the old Musalman nobility of the Province, being, in fact, the grandson of a very worthy merchant who amassed a considerable fortune, which he invested successfully in landed property near Dacca. The founder of the property was regarded with affection by the Hindus. He was very munificent and established hospitals and other charitable institutions, for which public services he was very properly awarded a title. The present Nawab probably perceives that if Dacca becomes capital of a new provincial government the enhanced value of his estate will be very appreciable... the Government of India recently... nominated him to the Viceroy's Legislative Council, as well as to the local Council. It is difficult to deny that his chief claim to these distinctions is that he has been a thick and thin supporter of the hated policy of Partition.[25]

Fraser's early excuse for taking away Dacca and Mymensingh to bring them into contact with 'a strong personal government' was thus abandoned in the heat of political expediency. The Viceroy once having taken a course of action was determined to see it through and scoffed at the hue and cry raised. While admitting that Dacca and Mymensingh were 'venting the air with piteous outcries', he, the great imperial overlord, 'had not found one single line of argument... nothing but rhetoric and declamation'. He dismissed the public protests as shriekings of children who had to be dealt with firmly and decisively. The opposition had to be viewed in relation to the vast interests at stake. The Viceroy was to do his duty, the vehemence of the agitation only spurred his determination not to relent. 'If we are weak to yield to their clamour now,' he warned the Secretary of State, '... you will be cementing and solidifying, on the eastern flank of India, a force already formidable and certain to be a source of increasing trouble in the future.'[26]

Curzon boasted to the adoring Mary, who was awaiting the birth of their third child in England, '... my speeches seem simply to have dumbfounded the opponents. The native newspapers are knocked silly and are left gasping.' Curzon went on to add, 'I showed that all the wild things that they said would ensue are pure fabrications.'[27]

The popular newspaper *Basumati* lamented the same week:

> If the Governor-General had regarded us as men of sense he would
> not have spoken as he has done. The same spirit runs through both
> his speeches (the Convocation speech and the Dacca speech), viz.,
> you are as children, you do not know what is good for you; all the
> same I shall do what I think is good for you.[28]

Interestingly, the Viceroy was proving to be equally high-
handed with his own people. In April Curzon sailed home for a
short sojourn; he had been granted a second term of viceroyalty
to finish the work he had begun in India and Lord Ampthill,
Governor of Madras, officiated as Viceroy in Curzon's absence.
On 17 June 1904, Under-Secretary A. C. Godley from the India
Office wrote in great exasperation to Ampthill about Curzon
saying, 'He seems almost to have lost sight of the merits of the
various questions, in which he has differed from the Cabinet or
from our Council (or they from him), and to be absorbed in a
struggle for prerogative, control, independence. In any of these
disputed matters, the thought that seems to rise in his mind is
that... "I have given my opinion, I have even reiterated it in two
or more despatches, I am the Viceroy of India, and, confound
you, how do you dare to set your opinion against mine?"'[29]

Meanwhile, in Curzon's absence, the two empowered civilians were having a field day: the chopping knives had come
out and through that long hot summer Bengal was further
divided. Fraser proposed a further transfer along communal
lines, giving the predominantly Muslim territory from north
Bengal to the new province.[30] But it was Risley, the Home
Secretary, who could be said to have drawn up the final partition plan. A scholar and a writer, author of *The Tribes and
Castes of Bengal*, Risley was not too popular with his colleagues.
The Lieutenant-Governor of Bengal, Sir John Woodburn is
said to have complained about Risley to Curzon over an alleged
indiscretion over the Calcutta Municipal Act of 1899.[31] But
Curzon had taken to Risley, making him a knight and elevating
him to the position of Home Secretary.

It was Risley who lulled the Viceroy into believing the agitation to be solely the work of a relatively small educated class with its 'organizing centre' in the old Sen capital of Bikrampur near Dacca. This region, said to supply one-third of the subordinate Indian officials in the government offices of Bengal, would stand to lose in the event of the partition of Bengal and was, therefore, hostile. Risley assured the Viceroy that 'they have had more than their share of appointments in the past and must be content for the future with what they can get.'[32]

When Congressmen complained that the proposed partition would lead to the loss of national unity, Risley told the Viceroy, 'Bengal united is a power; Bengal divided will pull several different ways. This is perfectly true and is one of the great merits of the scheme.'[33] Risley admitted that 'it is not altogether easy to reply in a despatch which is sure to be published without disclosing the fact that in this scheme as in the matter of amalgamation of Berar to the Central Provinces one of our main objects is to split up and thereby weaken a solid body of our opponents to our rule.'[34]

Risley could not have more blatantly outlined the imperial intentions of divide and rule. Enthused by these grandiose schemes Curzon forgot his own words to Balfour: 'I sometimes wonder whether 100 years hence we shall still be ruling India. There is slowly growing up a sort of a national feeling. As such it can never be wholly reconciled to an alien government...I believe a succession of two weak and rash viceroys could bring the whole machine toppling down.'[35]

Surendranath Banerjea mourned how the Viceroy was 'hopelessly out of touch with the spirit that his own reactionary policy had helped to foster.'[36] From western India, Gopal Krishna Gokhale issued a dire warning that this act of political repression would bring together all sections of public opinion, including those who had earlier kept themselves aloof from such agitations.[37] But busy trying to undermine the solidarity of the politically advanced Bengali, Curzon failed to see how he had touched a sensitive chord which was politicizing a vast mass of people. Stubbornly, the Viceroy clung to the belief that:

The agitation got up by the Native party to rest in the main depended upon the grossest misrepresentation. Hundreds of poor ignorant natives had been paid to hold up placards (frequently upside down) with English inscriptions printed upon them in Calcutta sent up by the gross before my arrival.[38]

While Curzon was in England the correspondence between him and Lord Ampthill,[89] shows that Curzon mentioned partition but rarely. The Viceroy probably wanted the shock of novelty to die down before he raked up the embers again.[40]

In the note of 6 December 1904, Risley justified his greatly expanded scheme by saying the Viceroy had considered it imperative to avoid a compromise scheme which 'would excite no opposition on part of the Bengalis and would commend itself to some other sections of public opinion'.[41] In saying this Risley seems to have implied that one criterion for acceptance was to provoke the Bengali. Curzon perhaps could not forget how his old mentor Strachey had detested Bengalis, exhorting that no position of importance be entrusted to them.[42] There is an element of vindictiveness in the attack, as though the Viceroy was impelled by a strong, perhaps subconscious impulse to strike.

Could some explanation be found in Mary's sudden and near fatal illness? As seen earlier, so many of the viceregal responses had been dictated by the emotional and physical settings of his life. In England in the summer of 1904, Mary had fallen gravely ill from an infection caught from the unsanitary plumbing at Walmer Castle where the family had taken up residence for a month. For days on end Mary had clung to life as by a bare thread. Distraught with fear and grief, Curzon had remained glued to her bedside. In India, Mary had been his only friend and confidante: to her he had brought his spoils of victory and she had offered him divine worship. The tensions of overwork and the unceasing pain in his back had been compounded by this new threat of losing Mary and through her the emotional anchor of his life. Had Curzon, close to being unhinged by panic, sought relief in hurting Bengal?

Besides, there had been the heady delight at having been able to push through unpopular reforms in the past. Curzon had come to believe that in India he could bulldoze his way through. He was prepared for the complaints, the petitions, the conference and the verbal fireworks, smugly expecting them to quieten as they had earlier done. Deluding himself into believing that the Bengalis were used to taking humiliations in their stride, he calculated this was one more for the mill.

Tragically, during his growing years, doting parents, teachers, and friends had made George Nathaniel Curzon feel that he was so superior that he had to carry the obligation of his admirers' ambitions. As he moved from Kedleston to Wixenford to Eton to Oxford, those surrounding him told him that he was an exceptional being. Neither his parents nor his early teachers had tried to protect him from the punitive anarchy of his own will. In their eyes he could do no wrong. At Eton, where there had been resistance there had also been the heady discovery that he could outwit it and stride on to his chosen goal. Conditioned, by and large, for adulation and success, he could no longer be at peace with himself unless he had ridden roughshod over opposition. He now dragged himself into self-destructive corners from which there could be no return.

Upon his return to India in December 1904, such was the Curzon's state that the only real flaw he saw in the Home Secretary's note of 6 December 1904 was in the presentation of the case. He had no dispute whatsoever with the scheme itself, even though it had been vastly expanded in his absence.[43]

When Sir Henry Cotton, President of the Congress and who had earlier been Curzon's chief commissioner in Assam, called on the Viceroy, he refused to meet him. Curzon boasted to Brodrick that the Congress had accepted defeat.[44] To an astonished Godley, Curzon said, everyone had agreed to the partition except the Congress, ' . . . who see in the sub-division of Bengal a weakening of Bengali influence in the future and a cruel postponement of the day when Cotton's ideal of an emancipated Bengal, under a Babu Lt.-Governor will be realised.'[45]

Lord Elgin had avoided open confrontation with the Congress although he had also declined to receive a deputation because it would convey a departure of policy. But Curzon plunged straight on, convinced he could do anything in India and get away with it. He was more concerned with making sure that the long and twisted route taken to arrive at the final scheme did 'not indicate either a change of position or a vacillation on our part, but a progressive advance towards the ideal which we always had in view'.[46] What was most important for Curzon was that his image as the strong and invincible Viceroy should not in any way be compromised.

In March 1905 he was claiming, 'Bombay is silent, Madras, though cogitating, is mute: nobody seems to take the faintest interest; and the Calcutta Congress, after thus exhausting one more damp squib from their pyro-technical armoury, will presently be sitting down to arrange for the eighth meeting to denounce the Viceroy...'[47]

Over 500 meetings had been held all over the province to condemn his action, yet Curzon stubbornly insisted on under-estimating public resentment. Three months later, when he gleefully reported, 'Conceive the howls! They will almost slay me in Bengal,'[48] it was as though the Viceroy was triumphantly demonstrating that 'supreme cheek' of the Eton boy who had dared keep a bar in his room. Carrying these infantile games into adult life, Curzon seemed to be now set on scoring off against the Bengali-speaking people who had dared to question British authority.

On 2 February 1905 when the Government of India had sent their proposals to Brodrick, the Secretary of State had written to say, 'It has been strongly urged here that the best way of giving relief to the Lieutenant-Governor, which all admit is necessary, would be to place portions of Bengal, probably Chota Nagpur and Orissa, under a Commissioner having a position like that of the Commissioner in Sind and invested, as may seem necessary with the powers of a Lieutenant-Governor.'[49] But Curzon wired back rejecting the Commissionership proposal. 'It would tend to consolidate the Bengali element by detaching it from outside factors and would produce the very effect that we desire to avoid. The best guarantee of the political advantage of our proposal is its dislike by the Congress party.'[50] He told Brodrick to go ahead ignoring the agitation in Calcutta which he claimed was:

> ...deliberately organised by a conspiracy between Wedderburn, Cotton, and 'Daily News' England, and instructed to get up meetings (which they did of schoolboys, petty agitators and Vakils) in the principal towns to denounce my so-called reactionary measures and to demand my recall.[51]

On 9 June 1905, Brodrick, giving his sanction, warned Curzon that he was underestimating the strength and substance inspiring the opposition. Brodrick cautioned:

> That a large and upon the whole homogeneous community of 41½ millions, with Calcutta as their Centre of culture and political and commercial life should object to the transfer of 3/5 of their number to a new administration with a distant Capital involving the severance of old and historic ties and the breaking up of racial unity, appears to me in *no* way surprising.[52]

Though Brodrick had vacillated initially, he had endorsed the Partition proposal. Later he was to try and absolve himself of the responsibility for the partition by implying that the Viceroy had hustled him into agreement. This was the impression Brodrick conveyed to Surendranath Banerjea whom he met in London in the summer of 1909. The Indian leader came away with the distinct impression that, '... but for the profound secrecy observed with regard to the final scheme, and our inaction owing to the absence of all information, the Partition of Bengal would not have been sanctioned by the Secretary of State. A timely deputation to England would have sealed its fate.'[53]

The summer of 1905 was a long, hot one in Bengal and the atmosphere was charged with tension. Curzon had written to Brodrick to say that the decision to partition should be announced and simultaneously executed. 'It is useless to attempt to persuade Bengal. The *fait accompli* is the only argument that will appeal to them...The more we say the greater will be the anger and commotion,' Curzon said.[54]

Highly inflammatory pamphlets were circulating in Calcutta and *mofussil* areas and the Government was aware of this.[55] Aurobindo's explosive *Bhavani Mandir* had found its clandestine way into many a home. From his retreat in Baroda, where he taught English, Aurobindo had not given up his efforts to prepare the country for an armed revolt. Identifying the Motherland with the avenging Bhavani, Aurobindo exhorted young men to give themselves up wholly, as he had, to the worship of the Divine Mother. In the twelve years since his return from England, the mystical, brooding Aurobindo was emerging as the high priest of a religious awakening. In 1902, he sent a young emissary Jatindra Nath Bandopadhyay to Calcutta to start a secret society. Jatindra Nath hired rooms at 108-C Upper Circular Road.[56] Eventually the group merged with the Anusilan Samity, with barrister Pramatha Nath Mitra as president and Chittaranjan Das, Aurobindo and Surendra Nath Tagore as office bearers.[57] Aurobindo spent the summer of 1902 in Calcutta making volunteers take vows with the Gita and the sword.[58] Nevertheless, the response was not encouraging. In 1904, Aurobindo's younger brother, Barindra, bemoaning two frustrating years in Bengal, had returned to Baroda a dejected man.[59] Had it not been for the Partition, the movement may well have fizzled out with a whimper.[60] On 19 July the final scheme for Partition was declared. 'The announcement fell like a bomb-shell upon an astonished public,' cried Surendranath Banerjea. 'We felt we had been insulted, humiliated and tricked. We felt...it was a deliberate blow aimed at the growing solidarity and self-consciousness of the Bengali speaking population...and to that close union between Hindus and Mohamedans upon which the prospects of Indian advancement so largely depended. For it was openly and officially given out that Eastern Bengal and Assam was to be a Mohamedan province.'[61]

Maharaja Jyotindra Mohan Tagore called a meeting at his palace in Pathuriaghata on the outskirts of Calcutta. Built in 1830 on the plan of a French architect, the 'Baithak-khana', as it was called, had handsome reception rooms, a well-stocked library, a banqueting hall and a billiard room. When the Tagores entertained, musicians played from the wrought iron turret overhead.[62] But this time the occasion was sombre. The meeting was to endorse the call for boycott of British goods. 'At the conference it was decided that the Maharaja should send a tele-gram to the Viceroy praying for a reconsideration of the orders passed, and urging that, if the Partition were unavoidable, owing to administrative reasons, the Bengali-speaking popula-tion should form part and parcel of the same administration.'[63]

Apart from Henry Cotton, there were other British sym-pathizers for the reversal of partition. While condemning the 'riotous rowdyism' of the agitators, the staid economic weekly *Capital* said, 'if this Partition of Bengal is repugnant to the feeling of the overwhelming majority of the people in the Province, the Government are unwise in forcing the measure on under such circumstances. I would appeal to the Governor-General in Council . . . not to rush the bill through the Council just yet. Give a little time. Sentiment has to be reckoned with everywhere. It largely rules the world. No Government can afford to flout the prevailing Public Opinion of a country.'[64]

The Editor of *The Stateman* had initially condemned parti-tion.[65] But waking up to the potentialities of the new weapon of *Swadeshi* and boycott, the paper was soon warning the Government, 'It has been apparent for some time past that the people of the province are learning other and more powerful methods of protest. The Government will recognize the new note of practicality which the present situation had brought into political agitation.'[66]

On 7 August, as the moist winds of the south-west monsoon swirled around the city, thousands of young men gathered at Calcutta's College Square to converge on the Town Hall. Holding high enormous black flags bearing messages, 'No Par-tition', 'United Bengal', '*Bande Mataram*', the youthful pro-cessionists flaunted yellow sashes tied around their heads like puggarees. This 'seething mass of humanity with black pennons floating overhead proceeded along College Street, Bow Bazar Street, Lal Bazar and reached their destination via Government Place.'[67] Excitement was running high, as long before the prescribed hour of five o'clock, the participants streamed out peacefully through the streets in rivers of purposeful, expectant humanity. At the Town Hall the crowds had been so large that they filled up the upper and the lower floors, and flooded the grand stairway with its marble pedestals, overflowing into the portico in the grounds below. One eyewitness said, 'The gathering was so immense that neither one, nor two, but three meetings had to be held... In the spacious hall upstairs and downstairs where two meetings were held and in the wings, the corridors, the vestibules attached to them were full with eager and expectant crowd... The scene, however, on the maidan outside baffles all description. The gathering was vast and immense and stretched as far as the eye could see.'[68] At one time organizers had resolved to drape the upper floor of the building in black, emblematic of the mournful occasion. An order for the black cloth had been given to the departmental store, Whiteway and Laidlaw until somebody remembered, in the nick of time, that the cloth was foreign.

Overnight emotions had risen to a fever pitch. People refused to participate in festivities where foreign salt or sugar were used. Priests declined to preside over ceremonies where foreign articles were offered. Students refused to appear in examinations where answers had to be written on foreign paper. So charged was the air with electricity that it became dangerous to appear outdoors wearing foreign clothing; a schoolboy who made this mistake narrowly escaped being lynched. A nationalist reported how his five year-old granddaughter returned a new pair of shoes simply because they were of foreign make.[69] The *swadeshi* spirit had come to assume a messianic fervour and Surendranath Banerjea recalls how once at a public meeting when he appealed to the members to take a *swadeshi* vow, 'a vast audience ... rose up with one impulse, and repeated in one voice the solemn words of the vow'.[70]

The giant was stirring and even the Viceroy was forced to admit 'the spectacle that has been presented by the streets of Calcutta during the past fortnight has not in my opinion been creditable to the Capital of a great Empire'.[71] But wreathed in his bluster he forced himself into believing that the agitation was 'a purely political movement organized by a small and disloyal factor on anti-British lines'.[72] He shut his eyes to the new mood of confidence sweeping across Asia: a mood which was focused by the Japanese triumph over Russia which had shattered the myth of European invincibility in the slaughter of the Tsarist fleet at Tsushima. The mendicant attitude of the older nationalists was being overtaken by an aggressive new militancy. The Viceroy failed to see that among the leading opponents of the partition were men like Sir Jyotindra Mohan Tagore, Sir Gurudas Bannerji, Raja Pearey Mohan Mukerji, Dr Rash Behari Ghosh, men who generally kept themselves aloof from ordinary political agitation and who had now come forward 'to oppose publicly the Partition'.[73]

October 16, the day of Partition, became a turning point in British India's history. The announcement of partition shattered the long, legendary belief in the eventual fairness of the British. It became a day of national mourning.[74] No food was cooked in Calcutta homes, the domestic hearths remained unlit. Shops and offices were closed and in several places newspapers were not distributed. Volunteers were out since the night before to look after the arrangements. Even before sunrise the streets of Calcutta began echoing with the cries of '*Bande Mataram*'. Thousands of sweltering, excited people marched barefoot to the riverside for a ritual self-purification bath. The stretch from the Nimtola Burning Ghat to the Howrah Bridge, a distance of nearly half a mile, was one long sea of faces. Poet Rabindranath's song '*Banglar mati Banglar jal*' thrilled patriotic sentiments to an all-time high and the ball had been set rolling for the mass *Raksha Bandhan* ceremony. Virtual strangers stopped each other on the streets to tie the *rakhi* thread, symbolizing brotherhood. The mood was the same everywhere, wild excitement, crowds filling the streets, shouting, weeping, cheering, embracing.

The city of Calcutta seemed to be paralysed by a general strike. Nationalist newspaper *Amrita Bazaar Patrika* reported:

Carters numbering about 11,000 struck work. Twelve Jute Press Factories, one Sugar Factory, one Shell Lac Factory, one Gun Factory and about seventy local mills were closed. Male and female hands working in these manufactories refused to work and consequently they had to be closed. The labourers in the Government Gun Factory had just commenced work when they heard the cry of 'Bande Mataram' and at once they left work and in a body came out and shouted Bande Mataram with thousands of others. There was not a single cart or a coolie at any of the four goods termini of the C. B. State Railway. All the marts and bazaars of Calcutta as well as Ultadangi, Talla, Ballygunge, Pattipukur, Belgatchin, Hatkhola, Shyambazar etc. were closed. All this showed distinctly that Bengal had been stirred to its innermost depths by an agitation unparalleled in the history of British rule.[75]

In a mood of great bitterness, the usually mild Gopal Krishna Gokhale compared the Curzon administration to that of the Moghul Aurangzèb. Scathingly Gokhale said:

There we find the same attempt at a rule excessively centralized and intensely personal, the same strenuous purpose, the same overpowering consciousness of duty, the same marvellous capacity for work, the same sense of loneliness, the same persistence in a policy of distrust and repression...[76]

While partition was carried out, the Viceroy was in Agra. On the evening of the fateful 16 October he dined with Sir James Latouche, Lieutenant-Governor of the United Provinces at the Agra Club.[77]

The gilded world of the British in India which seemed to be at high noon upon his arrival, suddenly appeared to stand dangerously close to the abyss, but the superior Viceroy failed to see how he had brought it there.

Chapter 17

A Humiliating Retreat

During his home leave, in the summer of 1904, Curzon had found that the British Cabinet were not prepared to adhere to his advice on Frontier problems. Over Afghanistan they had followed a vacillating attitude. Over the Tibet issue the British Cabinet first gave support, then withdrew it, arguing that the terms imposed upon the Tibetans were tantamount to annexation and far exceeded their instructions.[1]

What was more, while Curzon was in England, Kitchener had sent a memorandum to the Imperial Defence Committee, criticizing the dual system of military administration and threatening to resign if it was not abolished. By accident Curzon had come upon it. He recorded, 'The S. of S. subsequently informed me (Jan. 20, 1905) that the Memo had been handed over to the P. M. by Colonel Mulley, the Indian officer, who had been deputed to England to represent Lord K's views.'[2] What Curzon did not know was that Balfour, the Prime Minister, had invited Kitchener to present his views. Lady Salisbury had forwarded to Kitchener a 'Very Private' note written by her husband. It read, 'A. J. B. [Balfour] is very much concerned about the situation in India. He is much hampered because the information K. sends is secret, and he therefore cannot use it.'[3]

Curzon's friends had pleaded with him to stay back in England, arguing that the British Cabinet were made up of short-sighted men and that Curzon stood a chance of emerging as leader. Mary had, on an earlier visit in 1901, written to Curzon about the same matter. She then said:

> If you keep your health, as I pray God you will, you have the whole future of the Party in your hands. Arthur will not take the trouble to lead. St. John isn't inspiring enough, George Wyndham is a sentimentalist, and hasn't the hard sense to do strong things. So who is there but my Pappy? No one has anything like your vigour, and there is apathy in London about everything and everybody... They will need you to come back and wake them up. Great as your work is in India, it will be even greater in England, where the party is slipping down the hill of indifference and incapacity.[4]

Curzon could have stayed back. Mary's health and his own were sufficiently valid excuses. But there was a strange perversity in Curzon's nature which made him almost crave, as it were, for self-punishment. Downcast at having to leave Mary behind, he left knowing that he might not see her again until he finished his second term. The thought was unbearable, but he would not change his decision. The demon of self-destruction seemed to be propelling him.

Early in 1904 Mary had gone back to England for her third confinement. They had two daughters and had hoped for a son and even selected the name Irian-Dorian for him. However, it was again a girl. For once, Curzon, instead of wallowing in self-pity, only thought of Mary. He wrote to her with great tenderness:

> Darling, I felt how miserable you would be, and though of course I too was somewhat disappointed, I really felt it much more for you than for myself. However, I think we must entirely attribute it to me. You will remember that months ago we discussed and contemplated this, and that the name Naldera was arranged in consequence. So we will be content with our little Naldera and postpone Irian-Dorian till some future date. After all, what does sex matter after we are both of us gone?[5]

In fact Mary nearly died that summer. For five days and nights, as Mary hovered between life and death, Curzon had sat at her bedside, desolate and grief-stricken, writing down every word that she whispered. Curzon has recorded:

> I asked her whether in another world, if there were one, she would wait for me till I could come. 'Yes', she said, 'I will wait...' She asked that we might be buried side by side with a marble effigy of each of us looking towards each other, so that we might one day be reunited.[6]

By a miraculous chance, Mary did not die. By the end of October she was out of danger. But it was clear that she would never be perfectly well again, nor would it be safe for her to return to India. Nevertheless, he would not let this consideration stand in the way of what he believed was his great task in India. On the eve of his return to India he wrote to Mary:

> Amid all the great misery that we have been through shines out the consolation of many happy hours and tender moments and the memory of your beautiful and ineffaceable love. We have been drawn very close by this companionship in the furnace of affliction and I hope that it may leave me less selfish and more considerate in the future. To me you are everything and the sole thing in the world; and I go on existing in order to come back and try to make you happy.[7]

Back in Calcutta, Mary's empty rooms haunted him. Heartbroken and miserable, he wrote her, 'I have not dared to go into your rooms for I fear I would burst out crying... never in my life have I felt so forlorn and cast-down.'[8]

Unable to bear her husband's torment, Mary decided she must return. Her doctors did their best to dissuade her, warning her that she would not survive another Indian summer. But having made up her mind, nothing could dissuade her. Mary asked her brother-in-law Frank to send a telegram to Curzon informing him of her arrival. Curzon was delirious with joy.

> Was not yesterday the happiest day for years? For I got Frank's amazing telegram to say that you are actually coming... I could

hardly credit it and I went dancing off to Belvedere Ball (usually the most hateful of functions) in an almost indecent state of glee. I told everybody and they were all in a wild state of exultation. K[Kitchener] looked a new man and the room was one vast smile.[9]

Kitchener with his penchant for beautiful great ladies had gone out of his way to disarm Mary. When she returned to India in the spring of 1905, he invited her alone to his home for lunch, showing her his collection of priceless China.[10] Kitchener reminded Mary about her mother's collection of rare bottles, roguishly saying he hoped they were meant for him. A vastly diverted Mary declared, 'we greatly laughed at his greed'.[11]

In London, Balfour's Cabinet was in a fix. Brodrick wrote, explaining to Kitchener:

The difficulty lies in the fact that officials of almost every degree, including Lord Roberts, who are conversant with past working, adhere to what I call the dual control. They are fortified by what most people regard as the failure of successive attempts to improvise a better system.[12]

Though Lord Roberts had strongly supported the dual system, Balfour's Cabinet felt they could not ignore the claims of Kitchener. If Kitchener resigned, the British electorate could well withdraw its support to their shaky Cabinet. The situation was ominous. War clouds had gathered over the horizon, with skirmishes having broken out between Russia and Japan. The Commander-in-Chief had struck and drawn blood. He found the taste not unpleasant.

Before returning to India to resume his second term Curzon confessed:

I was aware that a severe struggle lay before me. I felt it a duty, however, to the Government of which I had been the head for so long not to desert it in the hour of trial but to sacrifice all personal consideration to the necessity of fighting its battles.[13]

When Curzon actually returned to India in December 1904, he had promised Balfour and Brodrick that he would examine the Military Member issue. At the Viceroy's Council meeting, Kitchener was found to be totally overruled by all the members.

Curzon argued:

> Administrative systems are not constructed to test exceptional
> men, but to be worked by average men...I believe that the com-
> bined duties which Lord Kitchener desires to vest in the head of
> the Army are beyond the capacity of any one man, of whatever
> energy, or powers.[14]

In his despatch Curzon drew attention of the British Cabinet
to the implications of vesting all authority in the Commander-
in-Chief:

> It would be similar to a situation in England if a Commander-in-
> Chief of the British Army possessed a seat in the Cabinet, if he
> were the sole representative of the army there, if he enjoyed the
> power of the rank of the Secretary of State for War in addition, and
> if His Majesty's Ministers were called upon to accept or reject his
> proposals with no dependent or qualified opinion to assist them.[15]

Kitchener merely added, 'I entirely dissent from the accompanying Despatch.'[16]

Kitchener in the meantime intensified his manipulation of
opinion in England. With the help of Major R. J. Marker,
Curzon's former ADC, now Private Secretary to the Secretary of
State for War, he published in *The Times* confidential documents
supporting his own case. When questioned, Kitchener flatly
denied having done so. Major Marker, still smarting over his
broken engagement to Mary's sister, had a vested interest in
humiliating the Curzons.[17] 'It was a Godsend you setting your
billet just in time to be of use to me,' Kitchener wrote to Marker.[18]
'I have told Gwynne of *The Standard* that if he appears to you,
you will show him confidentially the papers I sent you last mail
on the Military Administration.'[19] Kitchener had grumbled:

> Curzon told me the other day, after reading my papers, he still
> intended to support the Military Dept. for all he was worth. I told
> him that I feel it my duty to the army to resign on this question. He
> accepted this as the natural consequence. So I am preparing to
> pack up.[20]

In fact, Kitchener had stepped up his campaign at home,
aware that the Viceroy knew 'next to nothing of what is going

on at home' and was confident that 'Curzon and his Pocket Council will be [sic] difficulty to get over the Govt. at home'.[21]

Two days before the meeting of the Council in Calcutta in March 1905 Kitchener had despatched his version to General E. Stedman, an officer in the India Office. Curzon later wrote:

> So determined was he [Kitchener] that his version should be placed in the hands of the home authorities before the meeting of the Council at Calcutta. He sent from India to General Stedman a detailed and exhaustive reply to the minutes...in which the C-in-C freely attacked me behind my back. The letter covering 40 pages of typed foolscap was marked 'Private and Confidential'.[22]

The letter was later given to Colonel A'Court, military correspondent of *The Times*, who published an article, 'The Crises in India', on 30 May 1905.[23]

In June 1905 the government attempted a compromise, drafted by St. John Brodrick, but the tone of the despatch had been harsh and spiteful. The Military Department was to be retained. But the present Military Member, General Sir Edmund Elles—who had clashed with Kitchener—was to go.[24] He would be replaced by a junior officer in civilian clothes, renamed Military Supply Member, and would control only stores and transport. He would no longer give opinion on military questions nor veto the Commander-in-Chief's proposals.[25] Visibly upset, Curzon had rushed to Kitchener on 25 June and pleaded that the name Military Member be allowed to stay as of old and that the post be served by a soldier and not a civilian. Kitchener consented and a joint telegram to this effect was sent to the British Cabinet.[26]

Kitchener immediately regretted his mistake and confessed:

> I was rather upset at a slip of the tongue I made after a very heated discussion...after over an hour of a very straight talking from me he suddenly gave in and collapsed and in the excitement of the moment I said I would associate myself with him about the other demand.[27]

But having got Kitchener to agree, Curzon allowed himself to think that the victory was his and publicly criticized the

Home Government for having gone over the advice of the Viceroy's Council in making changes in the military administration of India. Sir Henry Fowler was provoked to put the following question in the House of Commons:

> I beg to ask the S. of S. [Secretary of State] India whether his attention has been called to the report of the speech contained in *The Times* today, delivered yesterday in the Viceregal Council by the Viceroy of India, in which the decision of H. M. G., with reference to the administration of the Indian Army and the Despatch of the S. of S. conveying that decision to the Viceroy are criticised, I might say severely, if not offensively.[28]

Kitchener gleefully wrote to Lady Salisbury, 'Curzon has, I think, given himself away by this very improper speech. I wonder what action the Government will take—he is evidently at their mercy.'[29]

Gullible and easily flattered, Mary had little idea of the two-timing game that Kitchener was playing with her husband. The Curzons were in Simla: the Viceroy had been ill for six weeks. In fact, only a month before Curzon was forced to resign, Mary was venting her spleen on Brodrick, accusing him of having 'embarked upon a policy of such insult and outrage' and claiming, 'There has never been such feeling in India as there is now against St. John—and the populace longs for his blood— and no one is more indignant than Lord Kitchener.'[30] Mary went on to add, 'there has never been any personal disagreement between Lord K. and George and they are on absolutely friendly terms...'[31]

Curzon was, however, by now aware that Kitchener had been sending secret telegrams to the British Cabinet. 'In the course of the Summer of 1905 at Simla a native gentleman of good position offered to place in my hands the cipher telegrams which Lord K was repeatedly sending in the war office code to Major M [Marker] in London,' he said.[32]

The time had come for him to resign.[33] The army's attempts to gain control over civilian rule was a popular issue with the British. But Curzon was not to leave on those grounds. When he did resign, it was over a paltry issue. Curzon suggested to

Brodrick the name of General Barrow as Military Supply
Member, saying that Kitchener had approved the choice.

> I had a long and confidential conversation with Lord K. He said
> he thought it the greatest pity that I would not get rid of Sir Elles
> and set General Barrow to work at once and that he himself
> thought General Barrow almost too good a man for the new billet.[34]

But as Curzon wrote, 'I heard at a later date that Lord
Kitchener after accepting General Barrow without a demur in
his conversations with me telegraphed to England behind my
back to say that he would not have him. This was told to me by
a Minister who had seen the telegram.'[35] Mary also said she
came to know that Kitchener '... saw Barrow before George
did & urged him to become the new Member and then denied
it,' now angrily declaring, 'I know him to be a liar'.[36] Curzon
should have now concentrated on exposing this deceit and having
Kitchener discredited rather than going on to insist upon
Barrow. But alas! Having taken a stand, foolish pride would
not permit him to withdraw.

Brodrick was getting impatient and testy and, Curzon
should have realized, was not going to allow him to have his
say. In fact, in a letter of 29 June Brodrick had warned, that
having, in Curzon's words, given way to Curzon twice before,
he was now going to have his say in nominating an officer for
the Military Supply Member from England.[37]

The telegram to the Viceroy on 1 August was particularly
censorious:

> We deeply regret the differences which have arisen between
> H.M.G. and yourself which have found public expression on
> your part, in your recent speech and in your Budget speech on
> March 29 ... both these speeches appear to us to be calculated to
> arouse ... public feeling in India. We cannot but feel that public
> opinion has also been inflamed by your intention to resign.[38]

Digging in his heels, Brodrick declined to consider the
appointment of General Barrow saying, 'This is also the view
of the Cabinet who are not willing to appoint Barrow. I hope
to telegraph you very shortly the name of the officer we propose

for M.S. [Military Supply] Dept.'[39] Three days later the Secretary of State telegraphed, 'I do not gather from your reference to Lord K. that he recommended General Barrow, but he knew your intention to recommend him . . . we cannot favour the selection of an officer who from the positions he has previously held can hardly be expected to inaugurate the new system with an open mind.'[40]

A headlong collision was in the offing and Curzon charged straight in:

> His Majesty's Government desire me to introduce a new system of military organisation into India. The only conditions upon which I can carry out their policy . . . are that I should receive their support, and be allowed the cooperation of the officer whom I consider best qualified for the purpose. If this is refused to me, I cannot accept any further responsibility for the discharge of the duty and a new Viceroy should be asked to attempt it.[41]

In the mistaken confidence that his resignation would not be accepted and that the threat would again get him his way Curzon sent off another telegram a week later:

> I am reluctantly driven to the conclusion that the policy of His Majesty's Government differs fundamentally from what I had thought had been agreed upon with the Government of India, and is based upon principles which I could not conscientiously carry into execution. In these circumstances my ability to act with advantage as head of the India Government has ceased to exist, and I beg you again to place my resignation in the Prime Minister's hands.[42]

Ten days later he received a royal telegram. The King said, 'With deep regret I have no other alternative but to accept your resignation at your urgent requests.'[43]

Curzon bitterly wrote to his father, 'I felt that I could not go on with honour or self-respect.'[44] But he had steered himself into this situation. As though not satisfied with humiliations already inflicted upon him, Curzon seemed to invite more by choosing to stay on in India for three more months in order to receive the Prince of Wales. Any other self-respecting man would have promptly made his exit. But Curzon wrote to the Prince of Wales, 'I own, I shall feel rather bitterly when I think of someone else doing the honours of Government House at Calcutta about which we have taken an enormous amount of trouble.'[45] Having planned the show, Curzon could not bear to give up presiding over the ceremonials, even when he ought to have known that to linger on after the farewells had been said could only be painful and self-humiliating. Curzon was both sick in body and mind, 'Our life here under St. John Brodrick has been a perfect hell,' Mary told her mother.[46] In one of her rare complaints against her husband, she had added:

> The great mistake which George made was in ever coming back to India. But as you know, nothing would prevent him. The strain of all this has been terrible ... I feel sometimes that I shall go out of my mind if have to bear much more stress and worry. The whole of my life seems sacrificed to this thankless public life.[47]

Mary was at the end of her tether. 'George is ill in bed from anxiety & prolonged diarrhoea and the children are all isolated ... on account of measles.'[48] Yet Curzon stubbornly hung on.

Brodrick had been ready with other humiliations for his old friend. As the next Viceroy, Lord Minto's date of arrival was uncertain, he wanted Lord Ampthill to take over as acting Viceroy and strip Curzon of his command. 'Brodrick has proposed our going at once,' Mary said bitterly, 'George's seven years of work are as though they had never been.'[49]

But the King intervened and his Secretary wrote to Curzon, 'You *are* to receive the Prince and Princess of Wales *officially* on their arrival at Bombay.'[50] Curzon gratefully acknowledged to the Prince, 'Owing to the ever memorable and considerate intervention of His Majesty the King, now the Viceroy is to have the great honour of receiving your Royal Highnesses in person.'[51]

The Curzons stayed on in Simla till October, answering thousands of letters of sympathy and sorrow at their premature departure. The Civil and Military service gave them a touching farewell.

In the meantime, on the plains below, 51 year-old civilian Sir Bampfylde Fuller, sporting a monocle in his right eye, had taken charge of the new province of Eastern Bengal and Assam. While recognizing that 'in Bengal it is the popular Lieutenant-Governors, like Woodburn and Bayley, rather than the strong and capable ones, like Mackenzie, Elliot and Campbell, who on the whole, do the most good,'[52] the Viceroy had perversely chosen his former Revenue Secretary, whom he had exasperatedly described as 'undoubtedly the least satisfactory Secretary whom I have had since I came to India, and I long for him to go.'[53] As could have been expected, the administration of Bampfylde Fuller was to be short-lived and disastrous: Fuller resigned on 3 August 1906, after barely ten months in power.

The new Lieutenant-Governor had set out by river to introduce himself to Dacca. As he neared the river wharf, he saw 'the housetops were black with masses of people ... and I was ceremoniously received by a crowd of leading citizens. But I noticed that there was no non-official Hindu gentleman amongst them.' The Lieutenant-Governor had raised his voice and boomed the greeting '*Mubarikbad*'. In reply, as he said, there was 'a deafening roar of welcome. But it was only from the Muhammadan voices. I found that the Hindus had already begun to show their resentment.' They chanted *Bande Mataram*. Clenching his teeth, Fuller said, 'I put a stop to this.'[54]

Already the Carlyle circular of 10 October banning the political participation of students was exciting emotions to a fever pitch.[55] In Calcutta, more than a thousand students had crowded into the home of Bengali ICS officer Charu Chandra Mallik on 27 October vowing 'not to bow down to the threats of the Carlyle circular.'[56] At a 2,000-strong rally of the Dawn Society on 5 November angry youngsters listened with rapt attention as orator after orator urged them to boycott the officialized University and its examinations.[57] On 9 November the day of the Prince of Wales's arrival in India, at another massive public meeting in Calcutta, 26 year-old Subodh Chandra gifted a sum of Rs 1 lakh for national education. So thrilled were the youthful audiences by the gesture that they unharnessed his carriage and dragged it all the way to his house.[58]

But taking a harsh pride in his work, Fuller had flown straight into the eye of the storm and issued what became known as '*Bande Mataram* circular'.[59]Next day, a settlement officer in Barisal town in Bakargunge district of Eastern Bengal was hounded by villagers armed with *lathis* and a District Magistrate was assaulted for interfering with the boycott of a steamer laden with European goods.[60] Fuller later said,

> Noisy crowds were formed which insulted European and Muhammadan passers-by, and in one case assaulted and beat the English manager of a local bank. They stopped carts that were laden with English cotton-goods and burnt their loads in mid-street. Processions were organized and marched about the towns with provocative cries of 'Bande Mataram'.[61]

Emotions were further aggravated when the tactless Fuller declared, 'I was like a man who was married to two wives, one a Hindu, the other a Muhammadan—both young and charming—but was forced into the arms of one of them by the rudeness of the other.'[62] When Bampfylde Fuller made plans to go to Calcutta for the Prince of Wales's visit he was warned that the agitators had sworn to hang a garland of old shoes round his neck.[63]

The Prince of Wales landed in Bombay as scheduled on 9 November. The Curzons had travelled down earlier in readiness to receive the royal visitors. Curzon had been 'quite ill' and barely able to move without help.[64] But when he had gone to receive the Prince it had been in full dress uniform, insisting that this was the only way for a Viceroy to greet his future monarch.

Mary had worn her 'most beautiful embroidered gown, with hat, parasol and gloves all to match'. To her 'horror' she found the Princess 'in a little common white Muslin . . . with a narrow blue sash and a little blue ribbon round her neck.' Appalled, Mary exclaimed, 'This for her great public entry to India'. There was no cheering from the crowds and Mary smugly put it down to the people being 'taken aback by their [the royal couple's] appearance,'[65] choosing to overlook the public anger over her husband's Partition of Bengal.

The royal couple had not taken too kindly to Curzon's reception for them. The Prince complained to his father how their rooms at Government House in Bombay had not been ready and upon their arrival the servants were still changing sheets. The King soothed his peevish son and daughter-in-law saying, 'it is simply inconceivable that Lord Curzon should have shown such bad manners. Let us hope that his state of health had much to do with it, as his mind before leaving India was simply unhinged.'[66] Brodrick's story of Curzon's suffering a nervous breakdown had obviously reached the King.

Curzon had been even more peremptory in the reception to his successor, childishly refusing to extend the courtesy due to an incoming Viceroy. He was not present at Government House, as protocol demanded, to receive Lord Minto when he reached Bombay.

Soon after receiving the Prince of Wales, Curzon fled to Agra to spend his last days in India by his beloved Taj Mahal. Seven o'clock in the morning would find the Viceroy dressed and ready on the site, happily supervising the restoration work. The pavilion in the palace of Shah Jehan had been joined together by brickwork for the purposes of an evening party and a ball in honour of the Prince of Wales's visit to India in 1876. Local talent had then been called in 'to reproduce the faded paintings on marble and plaster of the Moghul artists two-and-a-half centuries before'. The Viceroy had painstakingly laboured to erase this 'eyesore and a regret'.[67] From Agra the Curzons motored down to Akbar's old capital, Fatehpur Sikri, where the abodes of the Moghul emperor's Christian queen, Miriam, and favourite courtier Birbal, had been callously converted into a government guest-house. Banishing the travelling officials, Curzon had also ordered work to restore the monuments to their old glory.

The Viceroy had lingered in Agra as long as he could, wanting to memorize, as it were, the fascination and the poetry of the Taj Mahal that had so enthralled him when he had first seen it.

On 15 November he reluctantly boarded the train on the first lap of his journey back home, reaching Bombay the next day.

The Mintos landed in Bombay at six o'clock on the same evening, quietly driving down in the dusk to Government House. The Curzon children and their nannies had been packed off to the steamer to make room for the Mintos. But Curzon himself was not at hand. The Viceroy had been busy delivering his farewell address at the Byculla Club.[68]

When he had eventually acknowledged the Mintos, it was rather boorishly, clad in a casual coat and a pair of slippers— hardly the attire in which one Viceroy should receive another. But unable to cope with his humiliation, Curzon had characteristically retreated into schoolboy sulks, feeling that perhaps this was one way in which to compensate. Sir Walter Lawrence later wrote to Curzon:

> I am afraid that the reception of the Mintos' rankles & the story, exaggerated no doubt, has done harm in England. The story, as I heard it, was that you were not in the room at Govt. House to receive Lord Minto, that you came in some time after in smoking coat and slippers & after a few words said you wanted to talk to Dunlop-Smith. This, coupled with a very hugger-mugger swearing in the next day seems to have upset the new Viceroy & his wife & there has been a great deal of very bitter talk about it in India and at home... I don't suppose it matters much, but it may have helped to push Lord Minto into the camp of Lord Kitchener.[69]

But Curzon was past caring. He was hurt, broken and ill. The fight had gone out of him. His glorious hopes, his dreams, had come crumbling down. On 18 November, along with Mary, he sailed out of Bombay harbour never to return to India.

Curzon reached England on 3 December. As the train steamed in at Charing Cross, Mary could not help but wistfully note, 'old friends of a lifetime were mute'.[70] No member of the government came to meet them. The day after Curzon's arrival, the Conservative government collapsed and Balfour sent in his resignation to the King. A Liberal government came to power with Campbell-Bannerman as Prime Minister and John Morley as Secretary of State for India.[71] Curzon might well have remembered his own words on the eve of his departure for India seven years earlier:

> One goes out amid the glare of magnesium. How shall I return?[72]

His forebodings had proved prophetic.

When the King asked Balfour to submit Curzon's name for an earldom, Balfour said to Brodrick, 'I respectfully suggest the "waiting, waiting game" is the one to play. The pace is hot just now.'[73] The King had a moral stake in honouring Curzon for it was at his behest that Curzon maintained a silence over the controversy in India. A silence which could not have been easy. Apart from his own wounds, Mary was thirsting for Kitchener's blood. She had said, '...I feel no hesitation in telling what I know of his [Kitchener's] lies—and [letting] the King as well as everyone in England know what he is—& we shall ruin him if it is in our power...'[74]

The King had written to Curzon:

> I cannot but hope that on your return you may consider it advisable in the interests of the British Empire at large and especially as regards India, not to enter into any further controversy regarding the different issues with my Government which compelled you to resign...as the effect would be very serious. It is always undesirable, to wash one's dirty linen in public.[75]

Lawrence had reported that the King 'is also greatly concerned and anxious that when you come home you will meet your opponents with dignified silence.'[76]

It would have been a justifiable desire on Curzon's part to publish his version of the controversy on his return. He had intended to be heard and had assured his mother-in-law, 'There may be a few disclosures when I get home.'[77] Kitchener, alarmed at the prospect, had written to Major Marker begging him to see that the Viceroy was stopped:

> Curzon means to be as nasty as possible when he gets home. He means to attack me and is preparing by getting copies of a lot of secret notes that have passed between the Milty [Military] Dept. and the Headquarters...I think the press should be

warned of this . . . I hope you will be able to arrange for a douche of cold water coming on him from all around.[78]

Curzon respected his sovereign's wishes, Godley gratefully acknowledging that 'Curzon has promised to do nothing hastily'.[79] An earldom would have not merely mitigated some of his anguish at the humiliating termination of his Viceroyalty but also given Curzon a place in the House of Lords. But the new government under Liberal leader Campbell-Bannerman was not forthcoming, though as Viceroy of India Curzon was supposed to be above party politics.

Curzon never forgave his old friend Brodrick, saying he had been 'jockeyed out of office by a weak-kneed Cabinet and a vindictive Secretary of State' for the sake of an 'unscrupulous man without truth or honour'.[80] Upon Curzon's resignation being accepted, Mary said bitterly of Brodrick, 'you would not treat a dog as he has treated George'.[81]

Older by three years, Brodrick had been attracted to Curzon since his early Eton days, the time when young Curzon had barged by mistake into his railway carriage. Curzon later acknowledged, 'St. John Brodrick was a greater friend of mine at Balliol and in after life than at Eton'.[82] And when Curzon left for India, Brodrick said wistfully, 'You probably hardly know how much your departure means to me . . . It has been one of the brightest elements in my life to work with you and see you gaily flying the fences which I have laboriously climbed.'[83]

Over the years Brodrick had remained his admiring friend and well-wisher. Curzon's spinal problem troubled Brodrick and the latter's autobiography is scattered with his concern for Curzon's backache. 'Short of profligacy or alcoholism,' said Brodrick despairingly, 'I do not think any man could have done more than he did to shatter his health.'[84] Nor had Brodrick been grudging in his eloquence about Mary. When Mary, as Vicereine, visited London in 1901, Brodrick told Curzon,

I cannot let another mail go without telling you how great a success your dear wife has had in England—confirming some friendships, making many more, charming all. Everyone thinks she has developed so much in her interest in big things and the great problems without losing that delightful simplicity wh [ich] adds so much effect to her beauty and sweetness.[85]

Mary's mother, however, had not been taken in by Brodrick's concern or flattery. After the termination of his viceroyalty, Curzon acknowledged his mother-in-law's warning gratefully, for she had detected 'the animus of Brodrick who has been unstinting in his malevolence'.[86] Effectively camouflaged, the jealousy, seemingly almost non-existent, was a glowing coal, buried always deep and ready to flare up again.

Many years later St. John Brodrick published a sixteen-page printed note entitled *Relations of Lord Curzon as Viceroy of India with the British Government 1902–5*, adding to the title, 'seen and approved by the Earl of Balfour, June 1926'. In the pamphlet, Brodrick tried to imply that Curzon's viceroyalty had been terminated because the Viceroy was on the verge of a nervous collapse. Brodrick said:

> As early as 1902 his letters to his friends at home showed he was almost at the end of his tether ... As Viceroy he had no colleagues of similar experience to himself to consult. Lady Curzon, though possessing brilliant qualities, had not been brought up in England and had none of the traditional knowledge that many English-women possessed of the 'give and take' of public life. To a man thus seated on a pedestal, who felt himself standing up for the rights of the greatest dependency under the Crown against a body of men who, however, otherwise qualified, had most of them never set foot in India, ... it seemed imperative either to ride down all opposition or sacrifice his career in the attempt.[87]

Brodrick said Curzon's nervous exhaustion began two years after he had assumed his viceroyalty in India. It made him drive 'his officials remorselessly; he became intolerant of opposition, and it began to be whispered in India that to differ with the Viceroy was to imperil one's career.' Brodrick added that the Viceroy 'regarded the Council of India as mere registry office and the Secretary of State as the diplomatic representative of the Viceroy at the Court of St. James rather than as an individual responsible to Parliament for the Government of India.'[88]

The erstwhile Secretary of State waited till June 1926 to publish his version of the Viceroy's downfall in India. By then Curzon had been dead for over a year.

Ill-luck dogged Curzon on his return from India. Mary had rejoined her husband in India against the doctor's orders, bravely telling him, ' ... I would not have come unless I felt I had strength enough. I feel that the Indian heat could no more harm than the gnawing pain in my heart and my anxiety for you...'[89] The childlike Mary now showed that in a crisis she was capable of rising to great personal sacrifice. When she wrote to her husband she had not yet recovered from the illness which nearly killed her at Walmer Castle. She could barely move without support and spent most of her time in a bath-chair. Yet for the sake of her husband she was prepared to risk her life.

Mary did not die in India. But it was clear on their return to England in the bleak winter of 1905–6 that she had but a short time to live. To look after his wife and also to escape the 'unbroken ranks of hostility' of their old friends, her anxious husband took a house in Cap Martin on the Riviera. Mary continued to be gravely ill. Her heart gave problems: on climbing the stairs she had a blackout and suffered from suffocating breathlessness. But she bravely dismissed her troubles, saying only that 'it is such a nuisance'.[90]

In March the Curzons returned to London to at last take up residence in their own home, 1 Carlton House Terrace. But on 17 July Mary's health suddenly and seriously deteriorated and by evening it was clear that the end was not to be delayed. Doctors were hastily summoned and they worked steadily round the clock keeping her alive with oxygen, champagne and, finally, injections of strychnine. On the 19th afternoon, for a short moment it seemed that Mary was taking a turn for the better. She put up a fierce struggle. But it was of no avail. At 5.40 p.m. she collapsed. Curzon's arms were tightly clasped around her but she had gone for ever.

He came away from her deathbed shattered as never before. He put a flower and a photograph of himself in her hand, abiding by her request. Dazed, he wrote,

> There has gone from me the truest, most devoted, most unselfish, most beautiful and brilliant wife that a man ever had and I am left with three little motherless children and a broken life.[91]

With Mary's death Curzon felt he had lost the anchor of his life. He turned to Kedleston with his three infant daughters. As always his father, Revd Lord Scarsdale, was ready to receive him.

Epilogue

The partition of Bengal became a turning point in Indian political history. 'A cruel wrong has been inflicted on our Bengalee brethren and the whole country has been stirred to its deepest depth with sorrow and resentment as has never been the case before,' cried the usually moderate Gopal Krishna Gokhale in his presidential address to the Congress in December 1905.

Overnight a new spirit of patriotism filled the air. Bonfires were made of British goods, *Swadeshi* schools sprung up and indigenous companies began manufacture of *Swadeshi* goods. Hitherto uninvolved men, women and children poured out of their homes to lend support to the agitation. A protest petition from Eastern Bengal to the British Parliament was signed by 70,000 people.

By day, brave young men came out in the streets to picket bazaars and enforce the boycott. Companies of soldiers had been brought into Calcutta. Sentries paced up and down in front of public buildings and squadrons clattered down the boulevards. When there was a public meeting in a city square in north Calcutta, crowds of volunteers would march to the spot gustily shouting *'Bande Mataram'*. When sentries tried to ban the slogans, there were violent clashes. At night the streets were eerie and deserted, sometimes the flames of burning Manchester cloth could be seen glowing in the darkness.

By February 1906 the lean, intense and fanatical Aurobindo had returned to Calcutta to set into motion revolutionary terrorism on the Irish Sinn Fein lines. He became the leader of the extremist group in the Congress party, and the English-medium Bengali newspaper *Bande Mataram*, of which he was the guiding spirit, was soon demanding 'the absolute right of self-taxation, self-legislation and self-determination for the people of India'.

Aurobindo also took control of the inflammatory vernacular *Yugantar* founded by his younger brother Barindra in 1906. Equating revolution with religion, Aurobindo preached in so lofty and stirring a style that the Bengali youth flocked by the thousand into his fold. The circulation of *Yugantar* soared to 50,000. Overnight the towns and the countryside in Bengal were honeycombed with terrorist societies, the Dacca Anushilan alone boasting 500 branches.

In the new province of Eastern Bengal and Assam, the Lieutenant-Governor, Sir Bampfylde Fuller had come down with a heavy hand on the agitators. Ruthless police action broke up public meetings and even schoolchildren were not spared for singing national songs.

Events came to a head when peaceful delegates to a Bengal Congress conference in the town of Barisal were attacked with brute force and their leader Surendranath Banerjea arrested.

Shaken by the wave of unrest, Fuller had withdrawn the prohibitory orders banning the shouting of the slogan '*Bande Mataram*'. Nevertheless, three months later an embarrassed Government of India pressurized him to resign. But the memories of Bampfylde Fuller's repressive outrages were neither forgotten nor forgiven.

Surendranath Banerjea recalled how one night two Bengali youth came to him to say,

> ... We have formed a plan to shoot Sir Bampfylde Fuller; and we are going to—tonight for this purpose ... His Gurkhas stationed at Banaripara have been outraging some of our women, and we want to take revenge upon him.

As a counterblast to the Liberal Administration's declaration that the Partition of Bengal was 'a settled fact', political agitation was revived. On 21 July 1906, the day after a large protest meeting in Calcutta, the Indian employees of the East Indian Railway struck work all the way from Howrah to Bandel. The strike spread to Asansol and Jamalpur in Bihar where workers defied prohibitory orders to attend a political meeting.

Since the autumn of 1905, Calcutta and its environs had been in the grip of strikes. In September the printers and compositors in printing presses of the Government of India and the Bengal Secretariat had laid down their tools, to be followed by the clerks and the Indian assistants in the mammoth Burn Iron Works at Howrah. Soon work in the jute mills on the Hoogly was paralysed; there were strikes at the Clive Jute Mills and the Fort Gloucester Jute Mills. The railway strike of 1906 spread to the Eurasian railway guards and drivers. In February 1908 work in the Telegraph Office also began to be affected.

The partition agitation had not been confined to Bengal alone: by the end of 1905 the confidential reports of the Intelligence Branch of the Government of Bengal reported unrest in 23 districts in the United Provinces, 20 in the Punjab, 13 in the Madras Presidency, 24 in towns in the Bombay Presidency and 15 in the Central Provinces.

The Punjab was rent with cries of Arya for Aryans. The Punjabi Arya Samaj leader, Lala Lajpat Rai came to Calcutta to lend his weight to the Bengal struggle. Widespread agrarian riots broke out in Rawalpindi and Lahore. In 1907 the Government of India deported Lajpat Rai for his alleged involvement in the civil disobedience movement.

Maintaining close contacts with the extremist groups of Aurobindo and Bipin Chandra Pal, Tilak lent full-throated support to the calls for boycott and *Swadeshi*, providing ample echoes in Maharashtra. At the Calcutta Congress session of 1906, the extremists tried to elevate Tilak to the presidentship only to be foiled by the moderates. The split came in Surat the next year and the extremists found themselves squeezed out of the Congress for the next ten years. It needed Mahatma Gandhi to bring the extremists and Muslims back into the Congress mainstream again after World War I.

With each day, the tone of the Indian newspapers grew more shrill. When Lieutenant-Governor Andrew Fraser initiated prosecution against four Indian newspapers, *Yugantar* had retorted that no amount of prosecution could curb their spirit: those convicted became overnight heroes, a new printer being quickly registered after each conviction.

On 30 July 1907 the Government in desperation raided the offices of *Bande Mataram*. Though founded by Bipin Chandra Pal, the paper was largely run by Aurobindo. Tension ran high in the Court of Chief Presidency Magistrate D. H. Kingsford when Pal was produced as a witness in an attempt to convict Aurobindo. Pal refused to co-operate with the Bengal Government and was given a six-month sentence for contempt of court. When the sentence was read out, the crowds inside the Court went wild, shouting '*Bande Mataram*'. An infuriated Kingsford promptly arressted one slogan shouter, sixteen year-old Sushil Sen, sentencing him to fifteen strokes of the lash. As the whip lashed his young body, Sushil offered no resistance other than to cry out '*Bande Mataram*' with each stroke. The Secretary of State for India, Sir John Morley, was driven to condemn the savagery, saying Kingsford's floggings stink. And alarmed by the public resentment at the flogging, the Government transferred Kingsford to Muzaffarpur in Bihar.

On the moonless night of 30 April 1908 the terrorists struck in Muzaffarpur. The bomb was intended for Kingsford but by a tragic mistake it killed the wife and young daughter of Pringle Kennedy, a pleader at the Muzaffarpur Bar. Of the two young assassins, Prafull Chaki committed suicide, the other, Khudiram Bose was hanged to death but not before he had become a martyr. Students went into mourning for him. His photographs were sold in the market and young men took to wearing *dhotis* with his name woven as a border.

The British had not lost sight of the significance of the killings having taken place on *Amabasya* night, the auspicious dark night set aside for worship of Kali. The assassins had waited for twenty days in Muzaffarpur before throwing their bombs. Earlier, in a public speech, one speaker had advocated the sacrifice to Kali of white goats, which was now being interpretted as a sinister allusion to Europeans. Two days later the police descended upon a house in Maniktollah, on the outskirts of Calcutta, belonging to the family of the Ghosh brothers, and discovered quantities of crude bombs and seditious literature. Barindra was promptly arrested. Aurobindo was picked up the next day from 48 Grey Street in north Calcutta. Almost simultaneously Tilak was arrested in Maharashtra on charges of sedition: he had written two articles in *Kesari* in connection with the Muzaffarpur murders. Bombay mill workers went on large-scale strike in protest. From St. Petersburg, Lenin hailed the strike as the first political action of the Indian proletariat. Tilak was sentenced to six years rigorous imprisonment in Mandalay.

But the chain of political assassinations continued unabated. On the morning of 1 September 1908, two prisoners under trial for the Maniktollah conspiracy murdered Naren Goswami the government approver, in cold blood in front of the shocked inmates of Alipore Jail.

Lieutenant-Governor Andrew Fraser, considered one of the chief architects of partition, had for long been the natural target for the terrorists. In all, four attempts were made on his life. On three different occasions bombs were hurled at the trains on which he was travelling. Once at Narayanganj in Bengal the explosion shattered the railway track and the engine but Fraser himself was unhurt. On the fourth occasion, at a meeting at the YMCA hall in Calcutta, Fraser narrowly escaped being shot at point-blank range merely because, as the 18 year-old would-be assassin later explained, the trigger of his revolver got stuck.

On 11 December 1908, the Government of India revived Regulation III of 1818 for deportation without trial and summarily deported nine leaders. Included among the prisoners was Subodh Chandra Mallik. In Janaury 1909 several secret societies, including the Anushilan in Dacca were banned.

Aurobindo was still in Alipore Jail; his brother Barindra escaped the hangman's noose. Tilak was serving his sentence in Mandalay. In May 1909 Viceroy Minto gleefully reported that only hysterical students now manned the secret societies. Dispirited and leaderless, the revolution seemed to have ground to a halt.

Simultaneously, in a bid to mop up the support of the moderate nationalists, the Government announced constitutional concessions through the Councils Act of 1909, commonly known as the Morley–Minto reforms. In England, Curzon expressed alarm at proposed new Councils leading to the emergence of parliamentary government in India. Curzon said he was under the strong opinion that the Government of India would thereby become less paternal and less beneficent to poorer classes of the population.

The former Viceroy need not have worried. The Liberal Lord Morley confessed, 'If it could be said this chapter of reforms led directly or indirectly to the establishment of a parliamentary system in India, I for one would have nothing to do with it.'

The Partition of Bengal was revoked in 1911.

Curzon remained in the political wilderness in Britain until 1908 when he finally got his earldom and re-entered Parliament. Nevertheless, there had been stoic pleasure in that the world had abandoned him. 'Indeed I began to feel a sort of gloomy pride in my undistinguished distinction,' Curzon wrote. In the coalition government of 1915 he was Lord Privy Seal. When Asquith's Government fell next year, he became one of the four ministers in Lloyd George's War Cabinet. Simultaneously, he announced his engagement to the wealthy American widow, Grace Duggan, breaking off an eight-year liaison with the famous actress-novelist Elinor Glyn.

In Bonar Law's Cabinet, Curzon was the ostensible deputy Prime Minister. But on that crucial Whitsuntide weekend in 1923 when his succession was at stake, some death-wish seemed to have prompted him to go and bury himself in the country at remote Montacute and to await from there the royal summons to take up the high office which he felt was his. The summous had come, to tell him that Stanley Baldwin had been chosen instead.

The Davidson memorandum, which played a decisive role in destroying Curzon's hopes, today lies in the Royal Archives at Windsor with a Minute of the King's Secretary, Lord Stamfordham saying, 'this is the Memorandum handed to the King on Sunday, 20 May, [1923] and which Col. Walterhouse stated, practically expressed the views of Mr Bonar Law.' Significantly, Stamfordham makes no mention of Lord Davidson in his minute.

The Davidson Memorandum and Colonel Walterhouse's alleged use of it was said to be discovered by Bonar Law's biographer, Robert Blake, in 1955. In reply to a question by Blake, Davidson admitted that the memorandum had been prepared by him for Lord Stamfordham at the latter's request. Stamfordham knew then where the memorandum came from and whose mind it reflected. Stamfordham was therefore being less than honest when he said in his minute that it 'practically expressed the views of Mr Bonar Law'. The role of Lord Stamfordham in the story of succession thus is under a cloud.

There are further shadows. Curzon, when narrating the chain of events on that fateful Tuesday when he was not made Prime Minister, had said that 'Stamfordham's visit to me had been delayed to the hour when all protest or appeal from me was futile. For at that very hour—3.15—Baldwin was already at the palace, receiving his mission at the hands of the King.' Stamfordham, however, in his version says Baldwin's appointment was made only after his talk with Curzon. Stamfordham has recorded, 'I returned [after seeing Curzon] at once to Buckingham Palace and reported to the King what had passed at my interview with Lord Curzon.' His Majesty immediately 'afterwards saw Mr Baldwin and offered him the post of Prime Minister.'

Perhaps Stamfordham's bias against Curzon originated

twenty years before. The son of a Northumbrian parson, the outwardly unassuming Stamfordham had been the King's Secretary even while he was still heir to the throne. When the Prince of Wales landed in Bombay in November 1905, during the last days of Curzon's viceroyalty, Stamfordham was part of the royal entourage. He was therefore witness to Curzon's harum-scarum arrangements for the royal visitors at Government House, about which the Prince had complained to his father. Stamfordham also heard the tales of Curzon's churlish reception of his successor, Lord Minto.

During the four months of royal tour across India, Stamfordham probably had many other occasions to learn of the Curzonian arrogance. In army messes, embers had smouldered against Curzon's bold stand against the 9th Lancers. Incensed officers had filled the prince's ears with stories against the Viceroy. These momories perhaps still reverberated in Stamfordham's mind.

The 4th Marquess of Salisbury, son of the late Prime Minister who had given Curzon the Indian viceroyalty, had loyally travelled up from Devon in the milk train to endorse Curzon's claim to the premiership. Though he had to travel in the discomfort of a guard's van, Salisbury had worn formal dress—a frock coat and top hat—insisting that this was the only way for the Leader of the House of Lords to conduct official business.

On that Whitsuntide weekend Salisbury seemed to be Curzon's only friend. Arthur Balfour, whom Curzon affectionately referred to as King Arthur, was sick with an attack of phlebitis at a country house party in Norfolk. But when he came to hear of the succession issue, Balfour dragged himself from the bed to journey to London to pitilessly shoot down all chances of Curzon's elevation. Emphasizing that he was not referring to personalities, Balfour simply argued that the time had come when the Prime Minister must belong to the House of Commons. This reasoning automatically excluded Curzon, who was a peer.

Later, Lord Stamfordham was careful to record,

Lord Balfour said he was speaking regardless of the individuals in question, for whereas, on one side his opinion of Lord Curzon was based upon an intimate lifelong friendship, and the recognition of his exceptional qualifications, on the other, his knowledge of Lord Baldwin is slight and, so far, his public career has been more or less uneventful and without any signs of special gifts of exceptional ability.

Having delivered his blow, Balfour returned to his house party at Norfolk. Winston Churchill recounts how the ladies present had clustered around him to ask, 'And will dear George get the premiership?' 'No,' Balfour is said to have replied, 'dear George will not.'

Balfour's argument seems to have finally sealed Curzon's fate. It was not that Balfour liked Baldwin more but he certainly seems to have resented him less. Ironically, both Balfour and Curzon belonged to the same charmed circle—great ancestral homes, Eton, Oxford, the Souls had moulded them. However, Balfour had calmly watched Kitchener tearing Curzon to shreds in India. Now, with his argument to the King, Balfour had firmly pulled a shroud over Curzon's head.

In spite of his great wealth and his silken deprecatory manners, Balfour never married. According to Mary's biographer, Nigel Nicholson, for one long summer in 1901 Balfour was besotted by her. The lovely young Vicereine had been in London being lionized by society. Arthur Balfour, not yet Prime Minister, sought out his friend's wife, taking her on drives and picnics and to dinner until his steady lady-friend, Mary Elcho, had screamed in rage.

When Mary had playfully flaunted her conquest at her husband, Curzon had smugly said, 'Oh dear, it seems to me that you have fairly bowled over Master Arthur.' Pooh-poohing Balfour's ardour, Curzon had said deprecatingly, 'he is a tepid though delightful lover. So Pappy does not feel seriously afraid.' Perhaps Curzon should have.

In a handsome gentlemanly gesture Curzon had agreed to serve under Baldwin in 1923. 'I have every desire to retire,' he said, 'but as there are several things which, in the national interest, I ought to endeavour to carry through, and as my retirement at this moment might be thought to involve distrust of your administration.'[50] Curzon continued on at the Foreign Office.

Grace left on the trip to Argentina which she had told him she would cancel if he was made Prime Minister. Ill and in pain, Curzon went alone for a cure to France. 'I feel very much disheartened,' he said to Grace, 'because you do not take the slightest interest in what I say or do . . . You have never once asked me how was my leg. I told you I had crushed my nail in a door. . .' Grace replied, 'I am afraid no good will come of writing disagreeable letters to each other . . . the only possible thing is that we should lead our separate lives as much as possible.'

Baldwin's call for General Elections in early 1924 proved to be premature and disastrous for the Tories. Curzon lived to see the Labour Party in power and Ramsay Macdonald at 10 Downing Street.

When Baldwin returned to office for the second time in October, Curzon was given the paltry office of Lord Presidentship of the Council. He had taken it at Grace's insistence: Winston Churchill, his junior, was Chancellor of the Exchequer. All hope of political promotion now died in his breast, and with it, the hope of becoming Prime Minister and shaking the world. He was tired, lonely and ill. His heart was only at peace at Kedleston.

He begged Grace to go with him there, for Lord Scarsdale had died and Curzon had come into inheritance: 'All ask for Gracie and want to see the beautiful lady.' He wrote to her hopefully, 'One day you will take up your duties as Chatelaine of this place.' Even though Grace had told him, 'I would much rather not go at all—after all one's home is where one's heart is.'

At the end of February 1925 Curzon suddenly collapsed in the garden at Kedleston. For two days he rested in bed in the great state room wearing his overcoat and a pair of gloves—he abhorred radiators—under the large Adam canopy held up by gold palm trunks. Curzon had been scheduled to deliver a lecture in Cambridge in early March and characteristically he was not going to let this illness come in the way. When he stopped in London *en route*, it was clear that he was very ill. Grace later said:

> That night I was dining with Cecil and Alice Birgham, who had a big dinner-party from which we were all to go on to a Ball given by Lord Brassey. Towards the end of the dinner I was called to the telephone to answer an enquiry from the press—I was asked if George were ill, as there was a rumour that his speech would not be given. I said that I knew nothing of this ... but I did not phone to Cambridge because I was sure if George needed me I should have received a message from him ... So I went on with the Birghams to Lord Brassey's Ball.

The rumour was true. Curzon had collapsed again with a fatal haemorrhage while he was dressing for dinner. The end came peacefully on 20 March 1925. At least the struggle was over and he was free to be re-united with his beloved Mary in the family vault of the Memorial Chapel at Kedleston.

Curzon left behind the Victoria Memorial in Calcutta, his own Taj Mahal. The death of Victoria had offered Curzon the opportunity for the construction: £400,000 was raised, of which Bengal alone 'had given over £100,000', the *maidan* being selected as the site. Work did not begin till 1906, after Curzon had left India, and the building was only declared open in 1921. Curzon never saw it, though from England he had, with his exacting thoroughness, drawn up the design and the lay-out for the garden.

Curzon's statue has now gone from its commanding pedestal in the forecourt of the Victoria Memorial, where he had been the monarch of all he surveyed. It now lingers on at the back, in stately but somewhat forlorn majesty. The statue probably holds little meaning for India's younger generation. But the Victoria Memorial remains: grandiloquent, anachronistic, fashioned in the 'Italian Renaissance' style, yet unable to escape the Moghul touch, the shimmering white edifice floats in ethereal splendour over the vast, hot, expanse of Calcutta's *maidan*.

Robert de Courson, from Courson in Normandy (fl. 1066)

Giraline de Curcun

Richard de Curcun, 1st Lord of Kedleston (fl. 1135)

Robert de Curzon (d. c. 1205) = Alice Somerville

Stephen de Curcun, of Lockinge and Fauld

Richard de Curzon, ancestor of the Curzons of Croxall

Thomas de Curzon = Sibilla

Thomas de Curzon (d. 1245)

Richard de Curzon, 5th Lord of Kedleston (d. 1275)

Richard de Curzon (fl. 1297) = Joan

Ralph de Curzon

Richard de Curzon (fl. 1330) = Joan

Sir Roger de Curzon, Kt,

Sir John de Curzon, Kt 10th Lord of Kedleston (d. 1406) = Elenor, d. of Sir Robert de Twyford, Kt

John Curzon (b. 1394) = Margaret, d. of Sir Nicholas Montgomery, Kt

Richard Curzon (fl. 1432) = Mariora

John Curzon, 13th Lord of Kedleston (d. 1456) = Joan, d. of Sir John Bagot, Kt, of Blithfield, Staffs.

Richard Curzon, 14th Lord of Kedleston (d. 1496) = Alice, d. of Sir Robert Willoughby, Kt, of Wollaton. Notts.

John Curzon (d. 1512) = Elizabeth, d. of Stephen Eyre, of Hassop, Derbys.

Richard Curzon (1505–46) = Eleanor, d. of German Pole, of Radburne, Derbys.

John Curzon (d. s.p. 1549)

Francis Curzon (b. 1523) = Eleanor, d. of Thomas Vernon, of Stokesay, Salop.

John Curzon (1551–1632) = Millicent, d. of Ralph Sacheverall, of Stanton, Notts.

Sir John Curzon, 1st Bt. (1598–1686) = Patience, d. of Sir Thomas Crewe, Kt, of Stene, Northants.

Sir Nathaniel Curzon, 2nd Bt. (1635–1718) = Sarah, d. of William Penn, of Penn, Bucks.

Sir John Curzon, 3rd Bt. (1674–1727) d. unm.

Sir Nathaniel Curzon 4th Bt. (1675–1758) = Mary d. of Sir Ralph Assheton

Nathaniel, 1st Baron Scarsdale (1726–1804) = Lady Caroline Colyear, d. of 2nd Earl of Portmore

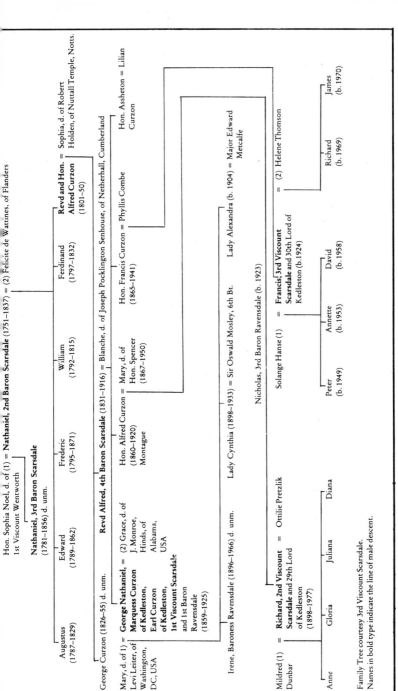

ANCESTRY OF CURZON OF KEDLESTON

Acknowledgements

I thank my husband Prafull whose idea it was in the first place and who has through out been my friend, philosopher and guide. He was to later teasingly refer to Curzon as 'the other man in my wife's life'! My two daughters, Brinda and Madhavi, for putting up with slapdash meals and neglected homework for seven years while their mother went in pursuit of Curzon. Without my family's support this book could never have been written. Now that it is finished it is theirs as much as mine.

Professor Hiren Chakrabarti (formerly Professor and Head of the Department of History, Presidency College, Calcutta, and currently Curator of Victoria Memorial, Calcutta), who listened to my half-baked idea in 1978 and originally guided me to prepare a Ph.D. thesis for Calcutta University. His wife, Mrs Gayatri Chakrabarti, I thank for her kind hospitality to me in London.

Professor Barun De, the first historian to give a patient hearing to my project, I thank deeply for recommending me to Professor Chakrabarti and for his continued interest in my progress. I am greatly indebted to Professors Ravindra Kumar, Purshottam Mehra, S. Gopal, Aparna Basu and Parthasarthi Gupta for their enlightened help. Any errors of fact or judgement in the book are wholly mine.

To psychiatrist Dr Ashit Sheth of Bombay I owe a particular debt of gratitude and affection for taking valuable time to help me psychoanalyse a man dead more than half a century. My father-in-law, Mr Dwarkadas Goradia I cannot thank sufficiently for kindly correcting the proofs for the umpteenth time, my mother, Mrs Savitri Daftary, for her unswerving faith in my ability to complete the project and my friends and family for patiently allowing themselves to be used as sounding boards. Mr S. Ghose-Chowdhury merits special mention.

The Librarians of the National Library, the libraries at the Bengal Club and Calcutta Club in Calcutta, the British Council Library, both in Calcutta and Delhi, the library at the Centre for Studies in Social Sciences, Calcutta, the India International Centre Library in Delhi, the British Museum, and the Eastbourne Central Library, I owe thanks for their boundless help. Mr R. N. Kapur for typing out the manuscript.

Dr Richard Bingle of the India Office Library and Records for his invaluable suggestions and occasional help in deciphering Curzon's hieroglyphics. Mr D. M. Blake, also of the India Office Library, for help in tracking down source material. Mrs M. Grey for checking references. Mrs Pauline Rohtagi, Ms Jill Spanner and R. Hamilton for researching photographs.

Mr Khushwant Singh for his kindness and patience in reading through the entire first draft and for putting me in touch with Mr Kenneth Rose.

Mr Kenneth Rose for giving me generously of his time, hospitality, a photograph of young Curzon and, most of all, the vital clue about the Paraman phenomena.

The third Viscount Scarsdale and Lady Scarsdale I owe a special debt of gratitude to for not merely their kind hospitality, but also for having personally helped me to track down specific material from the Kedleston Archives.

The 5th Marquess of Salisbury for kindly offering to let me have access to the papers of his ancestor, the 3rd Marquess of Salisbury.

Sir Anthony and Lady Susan Glyn in Paris for offering to tell me more about their grandmother's tempestuous love affair with Lord Curzon and to late Sir Brandon Rhys-Williams MP, for actually doing so.

Mr Nigel Nicolson I cannot thank sufficiently for not merely entertaining me at his famous home, Sissinghurst Castle, but also for generously putting at my disposal the Mary Curzon Papers and procuring Lady Alexandra Metcalfe's permission to use them.

Lady Alexandra Metcalfe for letting me quote from her parents' correspondence.

The publishers and the author would like to gratefully acknowledge the use of pictorial material from the following: Sri Aurobindo Ashram Trust, the Vice-Chancellor Burdwan University, the Madhuri Dhirajlal Desai Collection, Sunil Dutt, Messrs. Earnest Benn Ltd., Sir Anthony Glyn, The London Illustrated News Picture Library, India Office Library and Records, Messrs. Macmillan Publishing Co. Inc. (New York), Dr P.C. Mahtab, Lady Alexandra Metcalfe, Professor Nilmani Mukherjee, The National Trust, Nehru Memorial Museum and Library, Viscount Scarsdale, The Statesman Ltd., Umaid Bhavan Palace Museum Library, The Victoria Memorial.

Notes and References

ABBREVIATIONS

AP   Ampthill Papers.
CA. Curzon Additional Papers.
CP   Curzon Papers.
KMP Kitchener–Marker Papers.
MCP Mary Curzon Papers.
NNR Native Newspaper Reports.
OB   Oscar Browning Papers.
WLP Walter Lawrence Papers.

CHAPTER I

The Victorian Scene

1 Viscount Melbourne (1779–1848), twice Prime Minister of England.
2 Theo Aronson, *Victoria and Disraeli* (London: 1977), pp. 36–7.
3 Robert Wilson, *The Life and Times of Queen Victoria* (London: 1887), pp. 480–6.

4 Interview with the 3rd Viscount Scarsdale. Also see Kenneth Rose, *Superior Person* (London: 1969), p.18.

5 Benjamin Disraeli (1804–81), Prime Minister of Britain briefly in 1868 and again from 1874 to 1880.

6 William Ewart Gladstone (1809–98), four times Prime Minister of Britain between 1868 and 1892.

7 Jowett as quoted in Corelli Barnett, *The Collapse of British Power 1919–45* (London: 1972), p. 39.

8 Sir Harold Macmillan, *Winds of Change* (London: 1966), p. 41.

9 Philippa Pullar, *Gilded Butterflies* (London: 1978), p. 126.

10 Ibid.

11 Reminiscences by Curzon of His Early Life, CA 75.

12 Viscount Palmerston (1784–1865), twice Prime Minister of England. See R. C. K. Ensor, *England 1870–1914* (Oxford: 1936), pp. 137–43.

13 Calcutta, 12 February 1903, *Indian Speeches*, 4 vols. (Calcutta: 1900–6), vol. 3, p. 142.

14 Erik Erikson, *Childhood and Society* (London: 1974), p. 57.

15 Wayland Young, *Eros Denied* (London: 1964), p. 197.

16 Ibid., p. 203.

17 Philip Magnus, *Kitchener, Portrait of an Imperialist* (London: 1958), p. 132. Field Marshall Lord Kitchener of Khartoum (1850–1916), Commander-in-Chief in India, 1902–9, Secretary of State for War, 1914–16.

18 Letter of Sir J. Fitzjames Stephen, *The Times*, 1 March 1883 as quoted by Eric Stokes, *The English Utilitarians and India* (Oxford: 1959), p. 288. Sir James Fitzjames Stephen (1829–94), a Liberal and a Law Member of Viceroy Mayo's Council from 1869–72, author of *Liberty, Fraternity, Equality* (London: 1872).

19 Lord William Cavendish Bentinck (1774–1839), Governor-General of India 1833–5.

CHAPTER II

Kedleston Hall

1 All information on the family history and Kedleston Hall is drawn from interviews with 3rd Viscount Scarsdale, Curzon's *Kedleston Church* (privately printed, London: 1922) and Reminiscences by Curzon of his Early Life, CA 75.

2 Derby, 5 November 1898, *Indian Speeches*, 4 vols. (Calcutta: 1900–6), vol. I, p. xx.
3 See also, Kenneth Rose, *Superior Person* (London: 1969), pp. 14–19.
4 John Burke, *Dictionary of the Landed Gentry* (London: 1843).
5 Lawrence Stone, *The Family, Sex and Marriage in England, 1500–1800* (London: 1979), p. 327.
6 Ibid., p. 328.
7 Ibid., Richard Sheridan as quoted by Stone.
8 Ibid., p. 331.
9 Rose, op. cit., p. 18.
10 Curzon to Grace, 8 September 1921, CA 285. Mrs Grace Duggan, later Marchioness Curzon of Kedleston (1877–1958), married Curzon in 1917; daughter of Joseph Hinds, (who was formerly United States Ambassador to Brazil), widow of Brazillian millionaire Alfred Duggan.
11 Richard, too, had no sons, so on his death in 1977 the estate and title passed to the son of Curzon's younger brother Frank, Francis, who is today the 3rd Viscount Scarsdale, (see Family Tree).
12 Grace to Curzon, 29 September 1921, CA 66.
13 Grace to Curzon, undated [1922] CA 67.
14 Curzon to Grace, 6 December 1923, CA 287.
15 Nigel Nicolson, *Mary Curzon*, (London: 1977), p. 86.
16 Christopher Hibbert, *The Illustrated London News* (London: 1975), p. 16.
17 Interview with 3rd Viscount Scarsdale. The description of Kedleston is also based on the author's visits there in 1978 and 1984, the latter of which was made on the invitation of Lord Scarsdale.
18 Ibid.
19 Leonard Mosley, *Curzon, the End of an Epoch* (London: 1960), pp. 3–4.
20 Lord Riddell, *Lord Riddell's Intimate Diary of the Peace Conference and After, 1918–23* (London: 1933), p. 184.
21 Rose, op.cit., p. 26.
22 Baroness Ravensdale, *Little Innocents: Childhood Reminiscences* (London: 1932), p. 11. Baroness Irene Ravensdale (1896–1966).
23 Ibid., p. 10.
24 Sir Harold Nicolson, *Curzon, the Last Phase, 1919–1925* (London: 1934), p. 9.
25 Ravensdale, op.cit., pp. 11–12.
26 Ibid., p. 12.
27 Curzon to Farrar, 2 December 1882, CA 140.
28 Theo Aronson, *Victoria and Disraeli* (London: 1977), p. 84.

CHAPTER III

Imagined Tormentors

1 Leonard Mosley, *Curzon, the End of an Epoch* (London: 1960), p. 2.
2 Kenneth Rose, *Superior Person* (London: 1969), p. 18.
3 'Notes on Early Life', CA 20.
4 Lord Riddell, *Lord Riddell's Intimate Diary of the Peace Conference and After, 1918–23* (London: 1933), p. 184.
5 'Notes on Early Life', CA 20.
6 Among the Curzon Additional Papers at the India Office Library, I came across a 15-page hand-written paper by Blanche Scarsdale which makes nonsense of any theory that she was not intensely concerned with her eldest son. Blanche Scarsdale, 'About My Second Confinement when Baby George was Born, 1859' (a hand-written 15-page document), CA 190.
7 Ibid.
8 Lawrence Stone, *The Family, Sex and Marriage in England 1500–1800* (London: 1979), p. 270.
9 Ibid., pp. 270–2. Also see, Jonathan Gathorne- Hardy, *The Rise and Fall of the British Nanny* (London: 1972), pp. 33–57.
10 Mary Curzon, 'Journal' (in India), 1 February 1899, MCP.
11 See Mosley, op.cit., p. 2.
12 Blanche Scarsdale, note, op.cit.
13 Ibid.
14 Ibid.
15 'Notes on Early Life', CA 20.
16 'Reminiscences by Curzon of his Early Life', CA 75.
17 Ibid.
18 'Momentoes of Curzon's Early Childhood', CA 490/91.
19 Ibid.
20 Ibid.
21 Letters from Curzon to Members of his Family, CA 192.
22 Mosley, op.cit., p. 4.
23 Nigel Nicolson, *Mary Curzon* (London: 1977), pp. 85–6.
24 Oscar Browning (1837–1923), Eton Housemaster in Curzon's time, sacked. by Headmaster Hornby in 1875.
25 St. John Brodrick (1856–1942), later 1st Earl of and 9th Viscount Middle ton. First met Curzon at Eton in 1874, Secretary of State for India 1903–5.
26 Curzon to Browning, 4 January 1880, OB 1/446.
27 Curzon to Browning, 11 January 1877, OB 1/445.
28 See 'Reminiscences on Early Life', CA 75.

29 Ibid.
30 Curzon's Diary, 1866, CA 191.
31 Ibid.
32 'Reminiscences on Early Life', CA 75.
33 Alfred Curzon, Diary, by courtesy 3rd Viscount Scarsdale. Hon. Alfred Curzon (1860–1920).
34 'Reminiscences on Early Life', CA 75.
35 Nicolson, op.cit., p. 86
36 See Stone, op.cit., p. 293. Also see Gathorne-Hardy, op.cit., pp. 36–57.
37 Lord Scarsdale to Curzon, 2 December 1874, CA 142.
38 Ibid.
39 Curzon to Grace, 23 September 1916, CA 280.
40 Salisbury to Scarsdale, 10 November 1891, CA 270. 3rd Marquess of Salisbury (1830–1903), three times Prime Minister of Britain.
41 See Mosley, op.cit., pp. 56–7.
42 Ibid.
43 Ibid.
44 Ibid.
45 See Eric Erikson, *Childhood and Society* (London: 1974), pp. 43–107.
46 Curzon to Lady Scarsdale, 23 May 1869, CA 155.
47 Lord Scarsdale to Browning, 14 July 1874, CA 141.
48 Curzon to Lord Scarsdale, 12 November 1891, CA 270.
49 Mosley, op.cit., p. 72.
50 Nicolson, op.cit., p. 90.
51 Ibid., p. 98.
52 Curzon to Lord Scarsdale, 24 March 1904, CA 266/1–3.
53 Diary, 1866, op.cit.
54 'Notes on Early Life', op.cit.
55 Baroness Ravensdale, *Little Innocents: Childhood Reminiscences*, (London: 1932), p. 10.
56 William Churchill as quoted by Martin Gilbert, *Lady Randolph Churchill* (London: 1969), p. 93.
57 Ibid.
58 Ibid.
59 'Notes on Early Life', op.cit.
60 Gathorne-Hardy, op.cit., pp. 84–8.
61 'Notes on Early Life', op.cit.
62 Ibid.
63 Ibid.
64 Gathorne-Hardy, op.cit. p. 58.
65 Christopher Hibbert, *The Illustrated London News* (London: 1975), pp. 112–15.
66 Lawrence Stone, *The Family, Sex and Marriage in England 1500–1800* (London: 1979), p. 116.

67 Daniel Defoe as quoted by Stone, op.cit., pp. 244–5.
68 Ibid.
69 Ibid.
70 Mrs Philips to Lord and Lady Scarsdale, 11 June 1868, CA 155.
71 Mrs Phillips to Lord and Lady Scarsdale, 24 March 1869, CA 155.
72 Mary Powles to Lady Scarsdale, undated [1869], CA 155.
73 Curzon to Lady Scarsdale, 23 May 1869, CA 155.
74 Ibid., 11 February 1870, CA 156.
75 Ibid., 22 November 1871, CA 157.
76 Curzon to Lady Scarsdale, undated [1869], CA 155.
77 Curzon to Lord and Lady Scarsdale, 12 May 1869, CA 155.
78 Interview with 3rd Viscount Scarsdale.
79 Interview with Kenneth Rose. Mr Kenneth Rose first aroused my scepticism about the authenticity of the Paraman phenomenon when I met him in London in 1984.
80 'Notes on Early Life', op.cit.
81 Alfred Curzon, Diary, courtesy 3rd Viscount Scarsdale. See also Rose, op.cit., p. 21.
82 The notes are not dated but are handwritten on notepaper stamped with the address 1 Carlton House Terrace, Curzon's residence in London. Though Curzon took this house just before embarking on his viceroyalty, he came to occupy it only after his return, which suggests that the notes could not have been written before the termination of his viceroyalty.
83 Lord Riddell, *Lord Riddell's Intimate Diary of the Peace Conference and After* (London: 1933), p. 412.
84 Curzon to Mary, 28 May 1901, MCP.

CHAPTER IV

Golden Years at Wixenford and Eton

1 Curzon to Lord and Lady Scarsdale and grandmother, undated [May 1869], CA 192.
2 'Notes on Early Life', CA 20.
3 Mrs Phillips to Lord and Lady Scarsdale, 11 June 1868, CA 155.
4 'Notes on Early Life', CA 20.
5 Curzon to Lady Scarsdale, 12 May 1869, CA 155.

6 Revd Charles Kingsley as quoted in Ronald Hyam, *Britain's Imperial Century, 1815–1914* (London: 1976), pp. 82–3.
7 Ibid.
8 Ibid.
9 Curzon to Lord and Lady Scarsdale, 12 May 1869, CA 155.
10 Curzon to Lady Scarsdale, undated [1870], CA 156.
11 Ibid.
12 Powles to Lord Scarsdale, 30 July 1869, CA 155.
13 Ibid., 23 June 1870, CA 156.
14 Curzon to Lady Scarsdale, 5 May 1869, CA 155.
15 Lawrence Stone, *The Family, Sex and Marriage in England 1500–1800* (London: 1979), pp. 116–19.
16 Curzon to Lord and Lady Scarsdale, 12 May 1869, CA 155.
17 'Notes on Early Life', op.cit.
18 Stone, op.cit., p. 279–81.
19 Curzon to Lady Scarsdale, 11 November 1869, CA 155.
20 'Notes on Early Life', op.cit.
21 Ibid.
22 Curzon to Lady Scarsdale, 4 July 1869, CA 155.
23 'Notes on Early Life', CA 20.
24 Curzon, Diary, 1866, CA 191.
25 'Note-Book on First Journey Round the World', CP vol. 104.
26 The 5th Earl of Rosebery succeeded Gladstone as Prime Minister in 1894.
27 Arthur Balfour, Prime Minister from 1902–5.
28 Corelli Barnett, *The Collapse of British Power, 1919–1945* (London: 1972), p. 33.
29 Ibid., p. 34.
30 Mathew Arnold as quoted by Barnett, ibid.
31 'Notes on Early Life', CA 20.
32 Ibid.
33 Ibid.
34 Ibid.
35 Ibid.
36 Dod to Lord Scarsdale, 10 December 1872, CA 272.
37 Ibid., 19 March 1873, CA 272.
38 Madan to Dod, 13 March 1875, CA 272.
39 Dod to Lord Scarsdale, 19 March 1875, CA 272.
40 Curzon to Browning, 27 May 1877, OB 1/445.
41 Ibid., 9 July 1877.
42 Stone to Dod, 30 July 1877, CA 272.
43 Lord Scarsdale to Curzon, 2 December 1874, CA 142.
44 Curzon to Lord Scarsdale, 24 March 1873, CA 159.
45 Curzon to Lady Scarsdale, undated [1874], CA 160.

46 Kenneth Rose, *Superior Person*, (London: 1969), p. 41.
47 Curzon to Lady Scarsdale, 9 October 1873, CA 159.
48 Ibid., 5 October 1873.
49 Ibid., 9 October 1873.
50 Ibid., 21 January 1873, CA 159.
51 Ibid., 24 January 1873.
52 Lord Scarsdale to Curzon, 2 December 1874, CA 142.
53 Curzon to Lady Scarsdale, 20 October 1872, CA 158.
54 Lady Scarsdale to Curzon, 2 December 1874, CA 142. Lady Scarsdale probably meant 'taxing' when she said 'tacking'.
55 Dod to Lord Scarsdale, 27 July 1875, CA 272.
56 Curzon to Browning, 20 April 1878, OB 1/445.
57 Sir W. Lawrence, Diary, 10 February 1902, WLP vol. 27. Sir Walter Lawrence (1857–1940), Private Secretary to Curzon when he was Viceroy and author of, *The India We Served* (London: 1928).
58 Lord Scarsdale to Browning, 15 April [1878], OB 1/437.
59 Curzon, 'Eton Scrap Book', CA 290.
60 Curzon to Lord Scarsdale, 5 February 1874, CA 160.
61 Earl of Mayo (1822–72), Viceroy of India, 1869–72.
62 See S. Gopal, *British Policy in India 1858–1905*, (Cambridge: 1965), p. 101.
63 London, 29 October 1898, *Indian Speeches*, 4 Vols. (Calcutta: 1900–6), vol. I, p. v.
64 'Eton Scrap Book', CA 290.
65 Ibid.
66 Lord Scarsdale to Curzon, 4 June 1877, CA 142.
67 Rose, op.cit., pp. 37–8.
68 'Momentoes of Curzon's Early Childhood', CA 490/1.
69 Curzon to Brett, 9 October 1878, CA 144. Hon. Reginald Brett (1852–1934), Later Viscount Esher.
70 Paget Report, 'General Correspondence', CA 259.
71 Curzon to Browning, 6 October 1878, OB 1/445.
72 Lord Riddell, *Lord Riddell's Intimate Diary of the Peace Conference and After* (London: 1932), p. 411.
73 Ibid.
74 St. John Brodrick, *Relations of Lord Curzon as Viceroy of India with the British Government, 1902–5* (privately printed, London: 1926).
75 See, Havelock Ellis, *Studies in the Psychology of Sex*, 2 Vols. (New York: 1942), vol.1, part 2, p. 111. Ellis says: 'Derived from the name of an Australian novelist, Sacher-Masoch, Masochism is in the words of Kraft-Ebing: a peculiar perversion of the physical "vita sexualis" in which the individual affected, in sexual feeling and thought is controlled by the idea of being completely and unconditionally subject to the will of a person of the opposite sex, of being treated by this person as by a master, humiliated and abused. This idea is coloured by sexual feeling; the

masochist lives in fantacies in which he creates situations of this kind, and he often attempts to realize them'. Dr Ashit Sheth feels a vague and unpronounced form of masochistic tendency is fairly common and need not be tantamount to a sexual perversion. Also see, James C. Coleman, *Abnormal Psychology and Modern Life* (Bombay: 1981), p. 575. Coleman says: 'As in the case of the term "sadism", the meaning of "masochism" has been broadened beyond sexual connotations, so that it includes the deriving of pleasure from self-denial, expiatory physical suffering such as that of the religious flagellants, and hardships and sufferings in general.'

76 Interview with Dr Ashit Sheth, MD (psychiatry); also Havelock Ellis, *Studies in the Psychology of Sex*, 2 vols. (New York: 1942), vol. 1, part 2, p. 104–28.

77 Ibid. 'Sadism, as originally intended, denotes achievement of sexual stimulation and gratification through infliction of pain on a sexual partner. The term sadism derived from the name of Marquis de Sade (1740–1814) who for sexual purposes inflicted such pain on his victims as to be declared insane. The pain may be inflicted by such means as whipping, biting and pinching; the act may vary in intensity from light pats to severe mutila tion and in extreme cases to even murder. Now whereas the association between love and pain is said to exist amongst the most normal civilized men possessing well-developed sexual impulses, it is very easy for the sexual energy to pass over normal limits into the realm of cruelty.

78 Interview with Sir Brandon Rhys-Williams, MP, grandson of Mrs Elinor Glyn. Also see, Anthony Glyn, *Elinor Glyn* (London: 1968), pp. 226–8.

79 Glyn, ibid., p. 228.

80 Curzon to Grace, 19 July 1922, C.A. 286.

CHAPTER V

Eros Encountered

1 St. John Middleton, *Records and Reactions 1856–1939*, (London: 1939), pp. 25–6.

2 'Notes on Early Life', CA 20.

3 H.E. Wortham, *Victorian England and Cambridge* (London: 1956), p. 28.

4 Ibid., p. 32.

5 Ibid., p. 62.

6 Ibid., p. 97

7 See, Havelock Ellis, *Studies in the Psychology of Sex*, 2 vols., (New York: 1942), vol.1, part 4, pp. 1–65.

8 Ian Anstruther, *Oscar Browning* (London: 1983), p. 59.

9 Hon. Alfred Lyttelton, (1857–1913).

10 Hon. Edward Lyttelton, (1855–1942).

11 See Anstruther, op.cit., Chapters 4, 5 and 6.

12 Arthur Benson, *The Myrtle Bough* (Eton: 1903), p. 25.

13 See also, Ian Anstruther, *Oscar Browning* (London: 1983), p. 6. He says, 'Browning's account of the Eton disaster, told in his *Memoirs of Sixty Years*, related the facts without the background; so does his nephew, Hugo Wortham, who with the help of his Uncle's diary (now lost, perhaps destroyed) wrote his life in 1927.'

14 Curzon to Lady Scarsdale, 23 July 1873, CA 159.

15 Wortham, op.cit., p. 100.

16 Lyttelton to Curzon, undated, CA 152.

17 Lyttelton to Curzon, 27 July 1874, CA 152.

18 Farrar to Curzon, 11 January 1880, CA 148. Richard Farrar (1856–83).

19 Oscar Wilde to Curzon, 30 July 1885, CA 146A.

20 Ibid., November 1881.

21 Pater to Curzon, 13 May 1884, CA 146A. Walter Pater, along with Swin burne and Rossetti, was a high priest of the Aesthetic Movement.

22 Lyttelton to Curzon, 22 April 1884, CA 152.

23 Dod to Lord Scarsdale, 24 February 1876, CA 272.

24 Curzon to Lord Scarsdale, 25 February 1876, CA 266/1–3.

25 Wortham, op.cit., p. 101.

26 Ibid., p. 103–4.

27 Lord Scarsdale to Browning, 14 July 1874, CA 141.

28 Curzon to Browning, 20 July 1874, CA 141.

29 Wortham, op.cit., p. 300.

30 Curzon to Browning, 6 August 1874, CA 141.

31 Ibid., 23 August 1874.

32 Ibid., 7 January 1875.

33 Ibid., 5 May 1875.

34 Curzon to Browning, 15 February 1876, OB 1/445.

35 Dod to Scarsdale, 1 April 1876, CA 272.

36 Curzon to Lord Scarsdale, 5 December 1876, CA 266/1.

37 Lord Scarsdale to Browning, 26 December 1876, OB 1/437.

38 Ibid., 18 November 1877.

39 Ibid., January 1878.

40 Ibid., 22 July 1886.

41 Ibid., 26 December 1876.

42 Curzon to Browning, 2 February 1878, OB 1/445.

43 Ibid., 14 July 1874, CA 141.

44 Ibid., 26 March 1883, OB 1/447.

45 Ibid., 21 March 1879, OB 1/446.
46 Ibid., 27 June 1879.
47 Ibid., 4 October 1879.
48 Freud as quoted in Ellis, op.cit., pp. 80–1.
49 Curzon to Browning, 12 May 1883, OB 1/447.
50 Ibid., 3 January 1879, OB 1/446.
51 Curzon to Farrar, 29 January 1883, CA 140.
52 Ibid.
53 Brodrick to Curzon, 23 November 1879, CP vol.9.
54 Curzon to Farrar, 27 May 1883, CA 140.
55 Wortham, op.cit., 154–5.
56 Curzon to Farrar, 17 October 1880, CA 140.
57 Spring Rice to Curzon, March 1878, CA 263. Sir Cecil Spring-Rice (1889–1918).

CHAPTER VI

Oxford and Politics

1 Brodrick to Curzon, 29 May 1878, CP vol.9.
2 Lord Scarsdale to Browning, 18 November 1877, OB 1/437.
3 Brodrick to Curzon, 7 October 1878, CP vol.9.
4 Brodrick to Curzon, 16 May 1879, CP vol.9.
5 Kenneth Rose, *Superior Person* (London: 1969), p. 47.
6 Ibid., p. 48
7 Margot Asquith, *The Autobiography of Margot Asquith* (London: 1920), pp. 109–10.
8 Asquith, op.cit., p. 136.
9 Rose, op.cit., p.48.
10 Geoffrey Faber, *Jowett* (London: 1957), p. 21.
11 Jowett to Curzon, 31 December 1884, CA 145.
12 L. R. Johnson, as quoted by Ronaldshay, *The Life of Lord Curzon*, 3 vols. (London: 1928), vol. 1, pp. 42–3.
13 Farrar to Curzon, 11 January 1880, CA 148.
14 Curzon to Farrar, 28 May [1882], CA 140.
15 Lyttelton to Curzon, 4 July 1882, CA 151.
16 Lyttelton to Curzon, 4 July 1882, CA 152.
17 Lord Scarsdale to Curzon, 4 July 1882, CA 142.

18 J. W. Mackail as quoted by Ronaldshay, op.cit., pp. 58–9.
19 Curzon to Farrar, 27 May 1883, CA 140.
20 'Reminiscences by Curzon of his Early Life', CA 75.
21 Ibid.
22 Ibid.
23 Lyttelton to Curzon, 22 April 1884, CA 152.
24 Curzon, Diary, January to February 1886, CA 224.
25 *Cambridge Review*, 5 May 1880, as quoted by Ronaldshay, op.cit., pp. 45–6.
26 Ibid.
27 'The Conservatism of Young Oxford', *National Review*, June 1887, CP vol. 22.
28 Gladstone as quoted by S. Gopal, *British Policy in India, 1858–1905*, (Cambridge: 1965), p. 302.
29 Alan Bott, *Our Fathers* (London: 1931), p. 207.
30 Curzon to Brodrick, 12 November 1885, as quoted by Leonard Mosley, *Curzon* (London: 1960), p. 34.
31 Ibid.
32 Lord Vensittart, *The Mist Procession* (London: 1958), p. 232.
33 'Papers concerning Curzon's Campaign in S. Derbyshire and Southport, 1885–95', CA 208.
34 Curzon to Brodrick, 12 November 1885, as quoted by Mosley, op.cit., p.34.
35 'Curzon's Campaign in S. Derbyshire', op.cit.

CHAPTER VII

Tales of Travel

1 'Notebooks of First Journey Round the World, 1887–8', CP vol. 104.
2 'Notebook of Second Journey Round the World, 1892', CP vol. 105.
3 G. N. Curzon, *Problems of the Far East* (London: 1923), pp. 155–6.
4 Spring-Rice as quoted by Kenneth Rose, *Superior Person* (London: 1969), p. 248.
5 G. N. Curzon, *Tales of Travel* (London, 1923), pp. 231–6.
6 'Notebooks on First Journey', op.cit.
7 Ibid.

8 Earl of Ronaldshay, *The Life of Lord Curzon*, 3 vols. (London: 1928), vol. 1, p. 88.

9 Lyttelton to Curzon, 1 March 1890, CA 152.

10 See Mary Curzon, 'Journal', 1 February 1899, MCP.

11 London, 28 October 1898, *Indian Speeches*, 4 vols. (Calcutta: 1900–6), vol. 1, p. vii.

12 Ronaldshay, op.cit., p. 143.

13 *Tales of Travel*, op.cit., p. 41.

14 Ibid., p. 43.

15 Ibid.

16 Ibid., p. 52

17 G.N. Curzon, *Leaves from a Viceroy's Note-Book* (London: 1926), pp. 147–205.

18 See Curzon, *Tales of Travel*, op.cit., pp. 41–84.

19 Ibid., p. 67.

20 'Diaries and Notebooks of Journey to Persia', CA 277–8.

21 Curzon to Brodrick, as quoted by Leonard Mosley, *Curzon* (London: 1969), p. 38.

22 Ibid.

23 Ronald Hyam, *Britain's Imperial Century, 1815–1914* (London: 1976), pp. 136–7.

24 Curzon to Farrar, 24 March 1883, CA 140.

25 'Notebooks of Journey to Persia', op.cit.

26 'Notebook on First Journey', op.cit.

27 Ibid.

28 Ibid.

29 See Margot Asquith, *The Autobiography of Margot Asquith* (London: 1920), pp. 37–8.

30 Rose, op.cit., p. 82.

31 Ibid., p. 88.

32 'Notebook on First Journey', op.cit.

33 Ibid.

34 Ibid.

35 Ibid.

CHAPTER VIII

George and Mary

1 Max Egremont, *The Cousins* (London: 1977), p. 159.
2 Ronaldshay *The Life of Lord Curzon*, 3 vols. (London: 1928), vol. 1, pp. 41–2, for all information on origin of 'superior person' verse.
3 Lord Riddell, *Lord Riddell's Intimate Diary of the Peace Conference and After* (London: 1933), pp. 410–11.
4 Curzon to Grace, 18 November 1923, CA 287.
5 Brett to Curzon, 14 February 1878, CA 144.
6 Brodrick to Curzon, 7 April 1895, CP vol. 10.
7 Baroness Ravensdale, *Little Innocents: Childhood Reminiscences* (London: 1932), p. 9.
8 Lyttelton to Curzon, 7 June 1879, CA 151.
9 Margot Asquith to Curzon, 23 December 1888, CP vol. 12.
10 W. S. Blunt, as quoted by Ronaldshay, op.cit., p. 160.
11 'Reminiscences by Curzon of his Early Life', CA 75.
12 See Virginia Cowles, *Edward VII and His Circle* (London: 1956).
13 'Reminiscences', op.cit.
14 Leonard Mosley, *Curzon, the End of an Epoch* (London: 1960), p. 45.
15 Ribbelsdale to Curzon, 6 June 1885, CP vol. 11.
16 Asquith to Curzon, 10 May 1894, CP vol. 12. H.H. Asquith became Prime Minister of Great Britain from 1908–16.
17 Margot Asquith, *The Autobiography of Margot Asquith* (London: 1920), pp. 174–5.
18 Mosley, op.cit., p. 27.
19 Asquith to Curzon, 28 December 1892, CP vol. 12.
20 Lyttelton to Curzon, 19 April 1883, CA 151.
21 Ribbelsdale to Curzon, 19 December 1886, CP vol. 11.
22 Leonard Mosley, *Curzon, the End of an Epoch* (London: 1960), p. 47. The Mary Curzon papers to which I was granted access relate only to the years 1898–1906. I have had to therefore depend upon Curzon Papers and accounts given by Leonard Mosley, Nigel Nicolson, Kenneth Rose, Lord Ronaldshay and others for information on courtship and marriage of George and Mary.
23 Kenneth Rose, *Superior Person* (London: 1969), pp. 278–9.
24 See Nigel Nicolson, *Mary Curzon* (London: 1977), pp. 1–9.
25 Ibid., p. 38.
26 Ibid., pp. 39–40.
27 Ibid., p. 53.

28 Ibid.
29 'Notebook of Second Journey Round the World', CP vol. 105.
30 Mosley, op.cit., p. 48.
31 Nicolson, op.cit., p. 56.
32 Ibid., pp. 56–7.
33 Curzon to Mary, 4 March 1893, as quoted in Rose, op.cit., p. 280.
34 Mary to Curzon, 4 July 1894, MCP.
35 Curzon to Spring-Rice, 17 May 1893, CA 263.
36 Curzon to Mary, 3 September 1893, as quoted in Rose, op.cit., pp. 280–1.
37 Mosley, op.cit., pp. 56–7.
38 Ibid., p. 57.
39 Nicolson, op.cit., p. 72.
40 Ibid.
41 Ibid., p. 74.
42 Mosley, op.cit., p. 59.
43 Ibid., p. 58.
44 Nicolson, op.cit., p. 75.
45 Nicolson, op.cit., pp. 80–2.
46 Rose, op.cit., p. 289.
47 Nicolson, op.cit., p. 82.
48 Ibid., p. 89.
49 Ibid., p. 90.
50 Ibid., p. 103.
51 Curzon to Lord Scarsdale, 22 July 1898, CA 266/1–3.
52 Curzon to Salisbury, 18 April 1897, as quoted by Rose, op.cit., p. 322.
53 Mary to her parents, letter, undated, [1898], MCP.
54 Curzon, *Peers' Disabilities*, (London: 1894), CP vol. 19.
55 Lord Scarsdale to Curzon, 25 November 1898, CA 270.
56 Mary to her parents, undated [1898], MCP.
57 Ibid.

CHAPTER IX

Fin de siécle India

1 Ralph G. Martin, *Lady Randolph Churchill* (London: 1969), p. 195.
2 Sandhurst to Elgin, 19 June and 23 June 1897, as quoted by Stanley

A. Wolpert, *Tilak and Gokhale* (Los Angeles: 1962), p. 88.

3 P. L. Malhotra, *Administration of Lord Elgin 1894–99* (New Delhi: 1979), p. 108.

4 'Hitabadi', 11 Novembers 1896, as quoted by Malhotra, ibid., p. 111.

5 Hamilton to Elgin, 21 January 1897, ibid., p. 147. Earl of Elgin, (1849–1917), Viceroy of India 1894–9. Lord George Hamilton (1845–1927), Secretary of State for India, 1895–1903.

6 Elgin to Hamilton, 17 February 1897, ibid., p. 145.

7 Elgin to Hamilton, 4 April 1897, as quoted by Malhotra, ibid., p. 150.

8 Tilak in *Kesari*, 15 June 1897. Bal Gangadhar Tilak (1856–1920).

9 Autobiography of Damodar Chapekar as quoted in Wolpert, op.cit., pp. 88–9.

10 Ibid.

11 Curzon to Hamilton, 14 June 1899, CP vol. 158.

12 Ibid., 16 November 1899.

13 Valentine Chirol, *Indian Unrest* (London: 1910), pp. 48–9.

14 Gopal Krishna Gokhale (1866–1915), Member of the Viceroy's Legislative Council 1902–15, President of the Indian National Congress 1905.

15 See John R. McLane, *Indian Nationalism and the Early Congress*, (Princeton: 1977), pp. 44–9. See also, W. Wedderburn, *Alan Octavius Hume* (London: 1913), pp. 80.

16 W. C. Bonnerjee (1844–1906), President of the first session of the Indian National Congress, 1885 and 1892. Pherozeshah Mehta (1845–1915), President of the Congress in 1890. Surendranath Banerjea (1848–1925), entered the Indian Civil Service in 1871, was dismissed 1874, and was twice President of the Congress, 1895, 1902. Dinshaw Wacha (1844–1936), President of the Congress 1901. William Wedderburn (1836–1918), entered the Indian Civil Service in 1860, President of the Congress in 1889 and 1910, MP 1893–1900.

17 Dutt to Curzon, 15 September 1898, CP vol. 181. Romesh Chandra Dutt (1848–1909), passed the ICS exam in 1871. President of the Congress, 1899.

18 Curzon to Hamilton, 18 November 1900, CP 159.

19 Bipin Chandra Pal, *My Life and Times*, 2 vols. (Calcutta: 1932, 1951), vol. 1, p. 411.

20 Ibid., pp. 226–8.

21 Ibid., p. 253.

22 Hem Chandra Banerjee as quoted by Pal, ibid., pp. 256–7.

23 Ibid., p. 264.

24 As quoted in Amles Tripathi, *The Extremist Challenge* (Calcutta: 1962), p. 61. Swami Vivekanand (1863–1902).

25 Pal, op.cit., vol. 2, pp. 70–1.

26 Lansdowne to Kimberley, 22 August 1893, as quoted by Tripathi, op.cit., p. 69. Lord Lansdowne (1845–1927), Viceroy of India 1888–94.

27 Pal, op.cit., pp. 246–8.
28 Aurobindo Ghosh, *Bankimchandra* (Pondicherry:1954), p. 47. Aurobindo Ghosh (1872–1950).
29 Sumit Sarkar, *The Swadeshi Movement in Bengal 1903–8* (New Delhi: 1977), pp. 63–4.
30 Naoroji as quoted by Tripathi, op.cit., pp. 46–7. Dadabhai Naoroji (1825–1917), thrice President of the Congress, 1886, 1893, 1906, and author of *Poverty and Un-British Rule in India*.
31 Gorst to Lansdowne, 23 November 1888, as quoted in S. Gopal, *British Policy in India 1858–1905* (Cambridge: 1965), p. 181.
32 Curzon to Hamilton, 14 June 1899, CP vol. 158.

CHAPTER X

'We Might as Well be Monarchs'

1 Kenneth Rose, *Superior Person* (London: 1969), pp. 334–5.
2 *Indian Speeches*, 4 vols. (Calcutta: 1900–6), vol. 1, p. vii.
3 Curzon to Elgin, 17 November 1898, CA 182.
4 Mary to her father, 26 December [1898], MCP.
5 Ibid.
6 Ibid.
7 Mary to her parents, 4 January 1899, MCP.
8 Ibid.
9 Ibid.
10 Mary to her parents, 9 February 1900, MCP.
11 Mary to her parents, 4 January 1899, MCP.
12 See Earl of Ronaldshay, *The Life of Lord Curzon*, 3 vols. (London: 1928), vol. 2, pp. 23–4.
13 Mary to her parents, 4 January 1899, MCP.
14 St. John Brodrick, *Relations of Lord Curzon as Viceroy of India with the British Government, 1902–5* (London: 1926).
15 Mark Bence-Jones, *The Viceroys of India* (New Delhi: 1982), p. 160.
16 Ibid., pp. 160–1. Mand Lansdowne, wife of Marquess Lansdowne, Viceroy of India.
17 Mary to her family, 12 January 1899, MCP.
18 James Long, 'Echoes from Old Calcutta: its Localities', *Calcutta Review* (1852), vol. 36.

19 Ibid.
20 Ibid.
21 Ibid.
22 Pradip Sinha, *Calcutta in Urban History* (Calcutta: 1978), p. 13.
23 Ibid.
24 Ibid., p. 72.
25 Ibid., p. 79. Also see, Rajat Kanta Ray, *Social Conflict and Political Unrest in Bengal 1875–1927* (New Delhi: 1984), pp. 30–1.
26 Interview with Desmond Doig, Assistant Editor, *The Statesman*, Calcutta. Robert Clive (1725–74) first Governor of Bengal in 1757.
27 See Amalendu De, 'Raja Subodh Chandra Mallik and his Times', *Bengal Past and Present*, vol. 98, part 2, no. 187, July–December 1978.
28 Mary to her family, 12 January 1899, MCP.
29 Marchioness of Dufferin and Ava, *Our Viceregal Life in India*, 2 vols. (London: 1890), vol. 2, pp. 265–7.
30 Mary to her mother, 13 February [1899], MCP.
31 See *Englishman*, 15 to 31 December 1898.
32 Mary to her family, 27 January 1899, MCP.
33 Ibid.
34 Montague Massey, *Recollections of Calcutta* (Calcutta: 1918), p. 1.
35 Mary to her father, 13 February 1899, MCP.
36 John Beames, *Memories of a Bengal Civilian* (London: 1961), pp. 81–2.
37 Mary to her father, 12 February 1899, MCP.
38 Ibid., 10 February 1899.
39 Ibid.
40 See, *Englishman*, 20 December 1898 to 6 January 1899. Also, *Pioneer*, 20 December 1898 to 6 January 1899.
41 *Englishman*, op.cit. Sir Henry Cotton entered the Indian Civil Service in 1867, Chief Commissioner for Assam 1896–1902, President of the Indian National Congress 1904.
42 Curzon to Godley, 26 January 1899, CP vol. 158.
43 Nigel Nicolson, *Mary Curzon* (London: 1977), p. 84.
44 Elgin to Curzon, 31 August 1898, CA 182.
45 Mary to her family, 9 February 1900, MCP.
46 Nicolson, op.cit., p. 90.
47 Ibid.
48 Mary to her father, 17 January [1899] and Mary to her family, 29 January 1899, MCP. Irene was born in 1896 and Cynthia in 1898.
49 Curzon, *The British Government in India*, 2 vols. (London: 1925), vol. 2, p. 45.

CHAPTER XI

In Princely India

1 Mary to her parents, 4 January 1899, MCP.
2 See Sir William Barton, *The Princes of India* (London: 1934), and John Lord, *The Maharajas* (Delhi: 1971).
3 Bigge to Curzon, 13 January 1899, CA 85. Sir Arthur Bigge (1849–1931) was raised to the peerage as Lord Stamfordham in 1911.
4 Curzon to Mary, 4 November 1899, MCP.
5 Curzon to Hamilton, 25 October 1899, CP vol. 158.
6 Mary to Curzon, 13 November 1899, MCP.
7 See G. N. Curzon, *Leaves from a Viceroy's Note Book* (London: 1926), pp. 47–62.
8 Mary Curzon, 'Journal', MCP.
9 For greater details see Yvonne Fitzroy, *Courts and Camps in India* (London: 1926).
10 Mary to her parents, 10 December 1900, MCP.
11 Leonard Mosley, *Curzon* (London: 1960), p. 77.
12 Mary to her family, 24 November 1900, MCP.
13 Ibid.
14 Mary, *The Hyderabad Journal*, 1902, MCP.
15 John Lord, op.cit., p. 84.
16 *The Hyderabad Journal*, op.cit.
17 Mary to her father, 30 November [1899], MCP.
18 Ibid.
19 Curzon to Hamilton, 26 November 1899, CP vol. 158.
20 Mary Curzon, 'Journal', 1 and 2 December 1899, MCP.
21 Anthony Glyn, *Elinor Glyn* (London: 1968), p. 126. Also see John Lord, op.cit., p. 120.
22 Gwalior, 29 November 1899, *Indian Speeches*, 4 vols. (Calcutta: 1900–6), vol. 1, p. 168.
23 Curzon to the King Emperor, 19 June, 1901, CP vol. 135.
24 The Queen-Empress to Curzon, 13 September 1900, CP vol. 135.
25 Godley to Curzon, 22 November 1900, CP vol. 159. Sir Arthur Godley (1847–1932), Permanent Under-Secretary of State for India, 1883–1909.
26 Curzon to Hamilton, 19 June 1903, CP vol. 162.
27 Ibid., 29 August 1900, CP vol. 159.
28 Ibid.
29 Simla, 5 September 1902, *Indian Speeches*, 4 vols. (Calcutta: 1900–6), vol. 3, pp. 18–19.

30 Curzon to Hamilton, 9 January 1902, CP vol. 161.
31 Curzon to Hamilton, 26 October 1902, CP vol. 161.
32 For greater detail see Mortimer Menpes, *The Durbar* (London: 1903).
33 Mary to Curzon, as quoted in Nigel Nicolson, *Mary Curzon* (London: 1977), p. 167.
34 See Epilogue for the role played by Balfour in demolishing Curzon's chances of becoming Prime Minister.
35 D. C. Karve and D. V. Ambedkar (eds.), *Speeches and Writings of Gopal Krishna Gokhale*, 3 vols. (Poona: 1966), vol. 2, p. 190.
36 Hamilton to Curzon, 24 September 1902. CP vol. 161.
37 Curzon to Hamilton, 22 October 1902, CP vol. 161.
38 Ibid., 13 November 1902.
39 Curzon to Knollys, 15 November 1902, CP vol. 136.
40 Curzon to Balfour, 20 November 1902, CP vol. 161.
41 Ibid.
42 Balfour to Curzon, 12 December 1902, CP vol. 161.
43 'Notes on Early Life', CA 20.

CHAPTER XII

An Imperialist and a Gentleman

1 'Notes by the Viceroy in the Military Dept.', CP vol. 402.
2 Curzon to Hamilton, 13 June 1900, CP vol. 159.
3 'Military Notes', op.cit.
4 Curzon to Mary, 29 March 1901, MCP.
5 Curzon to Hamilton, 9 May 1900, CP vol. 159.
6 E. Maconochie, *Life in the Indian Civil Service* (London: 1926), p. 115–16
7 'Military Notes', op.cit.
8 Ibid.
9 An Anglo-Indian poet quoted in Denis Kinkaid, *British Social Life in India 1608–1937* (London: 1973), p. 242.
10 As quoted in Christopher Hibbert, *The Great Mutiny* (London: 1978), p. 39.
11 As quoted in Rajat Ray, *Social Conflict and Political Unrest in Bengal 1875–1927* (Delhi: 1984), p. 25. Rabindranath Tagore (1861–1941), poet, playwright, novelist, nationalist, son of Debendranath Tagore. Won the

Nobel Prize for Literature in 1913, knighthood conferred on him in 1915 which he renounced four years later.

12 Bipin Chandra Pal, *My Life and Times*, 2 vols. (Calcutta: 1932, 1951), vol. 1, p. 409.

13 Ibid.

14 Surendranath Banerjea, *A Nation in the Making* (Calcutta: 1963), p. 9.

15 Ibid.

16 Pal, op.cit., p. 149.

17 Ibid., p. 150.

18 Ibid., pp. 149–50.

19 Ibid.

20 See, Bampfylde Fuller, *Some Personal Experiences* (London: 1930), pp. 116–18.

21 Ibid., p. 118.

22 Pal. op.cit., p. 410.

23 Rivers Thompson as quoted in Ray, op.cit., p. 26.

24 Pal, op.cit., pp. 410–11.

25 'Military Notes', op.cit.

26 Bombay, 16 November 1905, *Indian Speeches*, 4 vols. (Calcutta: 1900–6), vol. 4, pp. 241–2.

27 Hamilton to Curzon, 3 January 1901, CP vol. 160.

28 See, Fuller, op.cit., pp. 119–20.

29 Ibid.

30 'Military Notes', op.cit.

31 Curzon to Lord Knollys, 14 December 1902, CP vol. 136.

32 'Military Notes', op.cit.

33 Curzon to Hamilton, 27 November 1902, CP vol. 161.

34 David Dilks, *Curzon in India*, 2 vols. (London: 1969), vol. 1, pp. 212–13

35 Curzon to Godley, 18 June 1902, CP vol. 161.

36 'Military Notes', op.cit.

37 Nigel Nicolson, *Mary Curzon* (London: 1977), p. 167.

38 Dilks, op.cit., vol. 1, pp. 202–3.

39 Curzon to Hamilton, 12 February 1903, CP vol. 162.

40 Curzon to Hamilton, 8 January 1903, CP vol. 162.

41 Maconochie, op.cit., pp. 115–16.

42 Curzon to Hamilton, 17 June 1903, CP vol. 162.

43 Calcutta, 7 February 1900, *Indian Speeches*, op.cit., p. 222.

44 Curzon to Godley, 7 December 1899, CP vol. 158.

45 Calcutta, 7 February 1900, *Indian Speeches*, 4 vols. (Calcutta: 1900–6), vol. 1, p. 216.

46 Ibid.

47 W. Lawrence, *The India We Served* (London: 1928), p. 235–6.

48 Jawaharlal Nehru, as quoted in Kenneth Rose, *Superior Person* (London: 1969), p. 339.

CHAPTER XIII

The Giant Awakes

1 This chapter is selective in its treatment of the Curzonian reforms, and does not claim to be exhaustive. As the study is of Curzon the man, the administrative reforms have been dealt with only in so far as they provide a better understanding of the human being.
2 Curzon to Hamilton, 11 October 1899, CP vol. 158.
3 Hamilton to Curzon, 10 February 1899, CP vol. 158.
4 D. G. Karve and D. V. Ambedkar (eds.), *Speeches and Writings of Gopal Krishna Gokhale*, 3 vols. (Poona: 1966), vol. 2, p. 330.
5 Curzon to Hamilton, 9 March 1899, CP vol. 158.
6 Curzon to Woodburn, 10 August 1899, CP 200.
7 Ibid.
8 Ibid.
9 Bright to Lawrence, 4 September 1899, CP vol. 200.
10 Ibid.
11 Curzon to Hamilton, 19 July 1899, CP vol. 158.
12 Curzon to Hamilton, 8 March 1899, CP vol. 158.
13 Ibid.
14 Curzon to Hamilton, 19 July 1899, CP vol. 158.
15 Curzon to Mary, 17 August 1901, MCP.
16 Hamilton to Curzon, 3 January 1901, CP vol. 160.
17 Mary to her father, 11 February 1899, MCP.
18 Calcutta, 11 February 1899, *Indian Speeches*, op.cit., p. 52.
19 Calcutta, 17 February 1900, ibid., p. 249.
20 Calcutta, 16 February 1901, ibid., vol. 2, pp. 206–8.
21 See 'Proceedings of the Education Conference, 1901', CP vol. 248. Except when specifically mentioned, the Viceroy's comments on Indian Education have been taken from this source.
22 See Sumit Sarkar, *The Swadeshi Movement in Bengal 1903–8* (New Delhi: 1977), p. 152.
23 See Amles Tripathi, *The Extremist Challenge* (Calcutta: 1962), p.54.
24 Curzon to Hamilton, 13 February 1902, CP vol. 161.
25 See Sarkar, op.cit., pp. 153–4.
26 'Education Conference', op. cit.
27 See Hamilton to Curzon, 19 September 1901, CP vol. 160.
28 Simla, 2 September 1901, *Indian Speeches*, 4 vols. (Calcutta: 1900–6), vol. 2, p. 311.
29 S. N. Banerjea, *A Nation in the Making* (Calcutta: 1963), p. 161.

30 Hamilton to Curzon, 5 January 1900, CP vol. 159.
31 Curzon to Hamilton, 10 September 1902, CP vol. 161.
32 Curzon to Dawkins, 24 January 1901, CP vol. 181.
33 Curzon to Brodrick, 3 March 1904, CP vol. 163.
34 Banerjea, op.cit., p. 161.
35 Amalendu De, 'Raja Subodh Chandra Mallik and his Times', *Bengal Past and Present*, vol. 98, part 2, no. 187 (July–December 1979), p. 34.
36 Ibid., p. 35.
37 Valentine Chirol, *Indian Unrest* (London: 1910), p. 85.
38 Pal, as quoted in Haridas and Uma Mukherjee, *Bipin Chandra Pal and India's Struggle for Swaraj* (Calcutta: 1958), pp. 12–13.
39 De, op.cit., p. 34.
40 Ibid., pp. 45–6.
41 Ibid., p. 48.
42 Ibid., p. 46.
43 Macpherson, Ch. Sec. Govt. of Bengal to Sec. Govt. of India, 6 April 1904, Home Dept. Public A, February 1905, Proceedings, pp. 155–67 (with enclosed memorials).
44 Curzon to Brodrick, 5 April 1904, CP vol. 1163.
45 Earl of Ronaldshay, *The Life of Lord Curzon*, 3 vols. (London: 1928), vol. 1, pp. 327–8.

CHAPTER XIV

'They Call Him Imperial George...'

1 Walter Lawrence, 'Diary', 1901, WLP vol. 26.
2 Ibid.
3 Ibid.
4 Ibid.
5 Ibid.
6 Mary to her family, 12 January 1899, and to her father, 17 January [1899], MCP.
7 Mary to Curzon, 9 June 1901, MCP.
8 Curzon to Mary, 23 July 1901, MCP.
9 Walter Lawrence, *India We Served* (London: 1928), p. 222.
10 Ibid., p. 238.

11 Baroness Ravensdale, *Little Innocents; Childhood Reminiscences* (London: 1932), p. 10.
12 Mary to her parents, 17 January [1899], MCP.
13 Mary to her mother, 11 January 1900, MCP.
14 Ibid., 3 May 1899.
15 Bampfylde Fuller, *Some Personal Experiences* (London: 1930), pp. 102–3.
16 Mary to her father, 13 February [1899], op.cit.
17 Ibid., Mary to her mother, 12 April 1899.
18 Ibid., Mary to her parents, 15 June [1897].
19 John Bradley, *Lady Curzon's India* (London: 1985), p. 39.
20 Mary to her father, 6 July [1899], op.cit.
21 Ibid., 21 June [1899].
22 Ibid., 8 February [1900].
23 Lawrence, 'Diary', op.cit.
24 Raymond Marker, as quoted by Peter King, *The Viceroy's Fall* (London: 1986), p. 98.
25 See Marker–Kitchener correspondence in Chapter XVII.
26 Mary to Curzon, 9 June 1901, MCP.
27 Curzon to Mary, 17 August 1901, MCP.
28 Lady Young to Curzon, 30 September 1901, CP vol. 230.
29 Young to Curzon, 30 September 1901, CP vol. 230.
30 Raleigh to Curzon, 22 September 1901, CP vol. 230.
31 Curzon to Young, 21 February 1902, CP vol. 230.
32 Young to Curzon, 27 February 1902, CP vol. 230.
33 Curzon to Hamilton, 21 May 1902, CP vol. 161.
34 Curzon to Hamilton, 3 May 1899, CP vol. 158.
35 Evan Maconochie, *Life in the Indian Civil Service* (London: 1926), p. 119.
36 Lawrence, *India We Served*, op.cit., pp. 224–5.
37 Curzon to Hamilton, 9 April 1902, CP vol. 161.
38 Mary to Curzon, 9 June 1901, MCP.
39 Lawrence, *India We Served*, op.cit., p. 227.
40 Brodrick to Curzon, 9 September 1900, CP vol. 10.
41 Bampfylde Fuller, *Some Personal Experiences* (London: 1930), p. 98.
42 Curzon to Mary, 23 July 1901, MCP.
43 Mary to her parents, 15 June [1899], MCP.
44 Curzon to Sir Schomberg Macdonnell, 25 July 1900, CP vol. 14. Hon. (Sir) 'Pom' Schomberg Macdonnell (1861–1915), Private Secretary to Lord Salisbury for fifteen years.
45 Fuller, op.cit., p. 96.
46 Margot Asquith, *The Autobiography of Margot Asquith* (London: 1920), p. 175.
47 Baroness Ravensdale, *Little Innocents, Childhood Reminiscences* (London: 1932), p. 11.

48 Curzon to Lord Scarsdale, 2 April 1901, CA 266/1–3.
49 W. Churchill, *Great Contemporaries* (London: 1937), pp. 277–8.
50 Lawrence, *India We Served*, op.cit., pp. 222–3.
51 Viscount D'Abernon, *An Ambassador of Peace* (London: 1929), pp. 48–51.
52 Lord Vansittart, *The Mist Procession* (London: 1958), p. 262.

CHAPTER XV

A Fateful Appointment

1 Rodd to Curzon, 12 December 1899, CP vol. 405.
2 .Hamilton to Curzon, 9 March 1900, CP vol. 159.
3 Sir Philip Magnus, *Kitchener, Portrait of an Imperialist* (London: 1958), p. 133.
4 Ibid.
5 Ibid., p. 152.
6 Ibid., pp. 1–5.
7 Ibid., p. 6.
8 Ibid.
9 Lady Cranborne, wife of Viscount Cranborne, became 4th Marchioness of Salisbury in 1903.
10 Magnus, op.cit., p. 148.
11 Mary to her mother, 9 March 1905, MCP and Magnus, op.cit., p. 147.
12 Earl of Ronaldshay, *The Life of Lord Curzon* (London: 1928), vol. 2, p. 109.
13 Curzon to Hamilton, 15 February 1900, CP vol. 159.
14 Curzon to Hamilton, 1 February 1900, CP vol. 159.
15 Ibid., 2 December 1902, CP vol. 161.
16 Ibid., 3 December 1902.
17 Magnus, op.cit., p. 202.
18 Curzon to Hamilton, 19 February 1903, CP vol. 162. Sir Edmund Elles, Military Member of the Viceroy's Council, 1901–5.
19 Curzon to Godley, 12 July 1899, CP vol. 158.
20 Curzon to Hamilton, 27 September 1899, CP vol. 158.
21 Hamilton to Curzon, 24 April 1903, CP vol. 162.
22 Magnus, op.cit., p. 206.
23 Ibid., p. 208.
24 Curzon to Hamilton, 7 May 1903, CP vol. 162.

25 Ibid.
26 Ibid., 9 July 1903.
27 'Private Correspondence relating to Military Administration in India: 1902–1905', CP vol. 412.
28 Magnus, op.cit., p. 209.
29 Miss Ella Christie, as quoted in Magnus, ibid., p. 205.
30 Macdonnell to Curzon, 4 October 1903, CP vol. 14.
31 Ibid., 29 December 1903.
32 Brodrick to Curzon, 17 December 1903, CP vol. 162.
33 Mary, 'The Persian Gulf Journal', 16 November [1903], MCP.
34 Curzon, *Tales of Travel* (London: 1923), pp. 247–50.
35 Ibid.
36 Ibid.
37 'Persian Gulf', op.cit.
38 Curzon to Frank, as quoted by Earl of Ronaldshay, *The Life of Lord Curzon*, 3 vols. (London: 1928), vol. 2, p. 303.
39 Brodrick to Kitchener, as quoted by Magnus, op.cit., p. 212.
40 Ibid., pp. 204–5.

CHAPTER XVI

The Partition of Bengal

1 Macaulay, 'Minute on Eduation, Feb. 2, 1835'. See Chapter 13.
2 Amles Tripathi, *The Extremist Challenge* (Calcutta: 1962), p. 86
3 See R. P. Cronin, *British Policy and Administration in Bengal* (Calcutta: 1977), p. 5.
4 Ibid.
5 Curzon to Hamilton, 30 April 1902, CP vol. 161.
6 Curzon, 'Notes', 24 May 1902, Home Department, Public A, December 1903, proceedings 149–60.
7 Sir Andrew Fraser (1848–1919), ICS, President of the Police Commission in 1901, Lieutenant-Governor of Bengal 1903–8.
8 H. H. Risley (1851–1911), ICS. Curzon appointed him Home Secretary to Government of India in 1902. Author of *The Tribes and Castes of Bengal* (1891).
9 Dutt to Curzon, 15 September 1898, CP vol. 181. Romesh Chandra Dutt

(1848–1909), joined the ICS in 1871, President of the Indian National Congress, 1899.

10 Curzôn to Hamilton, 27 September 1899, CP vol. 158.

11 Curzon to Hamilton, 3 January 1901, CP vol. 160.

12 Curzon, 'Note', 1 June 1903, Home Department, Public A, December 1903, Proceedings 149–60.

13 Curzon to Fraser, 24 July 1905, CP vol. 211.

14 Fraser, 'Note', 28 March 1903, Home Department Public A, December 1903, Proceedings 149–60.

15 Ibid.

16 Curzon, 'Note', 1 June 1903, op.cit.

17 Ibid.

18 *Sanjivani*, 28 January 1904, NNR.

19 S. N. Banerjea, *A Nation in the Making* (Calcutta: 1963), p. 171.

20 Ibid., p. 172.

21 Dacca, 8 February 1904. See Parliamentary Papers 1905, C. 2746, p. 222.

22 Curzon to MacDonnel, 1 June 1900, CP vol. 201.

23 Hewett, 'Note', 9 September 1904, Home Department Public A, February 1905, Proceedings 155–67. J. P. Hewett, ICS, Secretary to the Government of India, Home Department, 1895–1902. Chief Commissioner of Central Provinces, 1902–4, in charge of Indian Commerce and Industries Department, 1904–7.

24 Tripathi, op.cit., p. 97.

25 C. J. O'Donnell, *The Cause of Present Discontent in India* (London: 1908), p. 69.

26 Curzon to Brodrick, 17 February 1904, CP vol. 163.

27 Curzon to Mary, 22 February 1904, MCP.

28 *Basumati*, 27 February 1904, NNR.

29 Godley to Ampthill, 17 June 1904, AP vol. 14.

30 W. C. Macpherson to Government of India, 6 April 1904, Home Department Public A, February 1905, Proceedings 155–67.

31 Cronin, op.cit., p. 19.

32 Risley, 'Note', 6 December 1904, Home Department Public A, February 1905, Proceedings 155–67.

33 Ibid.

34 Ibid.

35 Curzon to Balfour, 31 March 1901, CP vol. 181.

36 Banerjea, op.cit., p. 172.

37 Gokhale, as quoted by D. G. Karve and D. V. Ambedkar (eds.), *Speeches and Writings of Gopal Krishna Gokhale*, 3 vols. (Poona: 1966), vol. 2, pp. 192–3.

38 Curzon to Brodrick, 23 February 1904, CP vol. 163.

39 Lord Ampthill, Governor of Madras, 1899–1906, Acting Viceroy 1904.

40 Curzon 'Note', 3 March 1904, Home Department Public A, February 1905, Proceedings 155–67.
41 Risely, 'Note', 6 December 1904, op.cit.
42 Tripathi, op.cit., p. 89.
43 Curzon, 'Notes', 25 December 1904, 6 January and 25 January 1905, Home Department Public A, February 1905, Proceedings 155–67.
44 Curzon to Brodrick, 29 December 1904, CP vol. 163.
45 Curzon to Godley, 5 January 1905, CP vol. 164.
46 Curzon, 'Note', 26 December 1904, op.cit.
47 Curzon to Godley, 16 March 1905, CP vol. 164.
48 Curzon to Ampthill, 8 June 1905, CP vol. 210.
49 Brodrick to Curzon, 20 May 1905, CP vol. 175.
50 Curzon to Brodrick, telegram, 24 May 1905, CP vol. 175.
51 Curzon to Brodrick, 1 June 1905, CP vol. 164.
52 Brodrick to Curzon, despatch of 9 June 1905, Home Department Public A, October 1905, Proceedings 163–98.
53 Banerjea, op.cit., p. 173.
54 Curzon, 'Note', 24 June 1905, Home Department Public A, September 1905, Proceedings 302.
55 'Report on Agitation Against Partition of Bengal', Home Department Public A, June 1906.
56 Amalendu De, 'Raja Subodh Chandra Mallik and his Times', *Bengal Past and Present*, vol. 98, part 2, no. 187 (Calcutta: July–December 1979), p. 43. See also, Sumit Sarkar, *Swadeshi Movement in Bengal: 1903–1908*, (New Delhi: 1977), pp. 465–83.
57 De, ibid.
58 Sarkar, op.cit., p. 470.
59 Ibid., p. 473.
60 Aurobindo Ghosh, *Sri Aurobindo on Himself and on the Mother* (Pondicherry: 1953), p. 41.
61 Banerjea, op.cit., p. 173.
62 See Desmond Doig, *Calcutta, An Artist's Impressions*, (Calcutta: n.d.).
63 Banerjea, op.cit., p. 173–4.
64 *Capital*, 7 September 1905.
65 Banerjea, op.cit., pp. 173–4.
66 Ibid., p. 180.
67 *Amrita Bazar Patrika*, 8 August 1905.
68 Ibid.
69 Banerjea, op.cit., pp. 176–92.
70 Ibid., p. 212.
71 Curzon, 'Note', 3 October 1905, Home Department Public B, October 1905, Proceedings 114–15.
72 Curzon to Brodrick, 9 October 1905, Home Department Public A, September 1905, Proceedings 302.

73 Gokhale, op.cit., p. 192.
74 See Banerjea, op.cit., pp. 197–203.
75 *Amrita Bazar Patrika*, 17 October 1905.
76 Gokhale, op.cit., p. 188.
77 *Amrita Bazar Patrika*, 17 October 1905.

CHAPTER XVII

A Humiliating Retreat

1 For greater detail see, P. Mehra, *Younghusband Expedition* (Bombay: 1968).
2 'Private Correspondence relating to Military Administration in India: 1902–1905', CP vol. 412.
3 Philip Magnus, *Kitchener, Portrait of an Imperialist* (London: 1958), p. 207.
4 Mary to Curzon, 5 July 1901, MCP.
5 Nigel Nicolson, *Mary Curzon* (London: 1977), p. 172.
6 Ibid., p. 180.
7 Curzon to Mary, 25 November 1904, MCP.
8 Ibid., 13 December 1904.
9 Ibid., 21 January 1905.
10 Mary to her mother, 9 March [1905], MCP.
11 Ibid.
12 Magnus, op.cit., p. 212.
13 'Military Administration', op.cit.
14 Ibid.
15 Ibid.
16 Despatch 36, with Minute of Dissent by the Commander- in-Chief, 23 March 1905, Military Department Proceedings, 1905.
17 Nicolson, op.cit., p. 200.
18 Kitchener to Marker, 15 February 1905, KMP.
19 Ibid., 12 January 1905.
20 Ibid., 19 January 1905.
21 Ibid., 17 January 1905.
22 'Military Administration', op.cit.
23 Ibid.
24 Military (Secret) Despatch no. 67 from the Secretary of State, 31 May 1905, Military Department Proceedings, 1905.

25 Ibid., Despatch 66.
26 Curzon to Balfour, 26 June 1905, CP vol. 175.
27 Kitchener to Marker, 6 July 1905, KMP.
28 'Military Administration', op.cit.
29 Magnus, op.cit., p. 221.
30 Mary to her mother, 27 July [1905], MCP.
31 Ibid.
32 'Military Administration', op.cit.
33 Curzon to Ampthill, 12 May 1905, AP vol. 18.
34 'Military Administration', op.cit.
35 Ibid.
36 Mary to her mother, 2 October [1905], MCP.
37 Brodrick to Curzon, 29 June 1905, CP vol. 164.
38 Brodrick to Curzon, 1 August 1905, CP vol. 195.
39 Ibid.
40 Ibid., 4 August 1905, CP vol. 175.
41 Curzon to Brodrick, 5 August 1905, CP vol. 175.
42 Curzon to Brodrick, 12 August 1905, CP vol. 175.
43 The King-Emperor to Curzon, 22 August 1905, CP vol. 136.
44 Curzon to Lord Scarsdale, 17 August 1905, CA 266/1–3.
45 Curzon to the Prince of Wales, 23 August 1905, CP vol. 216.
46 Mary to her mother, [August 1905], MCP.
47 Ibid., 24 August [1905].
48 Ibid., [August 1905].
49 Ibid., 24 August [1905].
50 Lord Knollys to Curzon, 21 September 1905, CP vol. 136.
51 Curzon to the Prince of Wales, 5 October 1905, CP vol. 216.
52 Curzon to Hamilton, 10 August 1902, CP vol. 161.
53 Curzon to Ibbertson, 23 December 1901, CP vol. 230.
54 Bampfylde Fuller, *Some Personal Reminiscences* (London: 1930), pp. 125–6.
55 Bengal Circular no. 1679, P. D., 10 October 1905, Home Dept. Public A June 1906, Proceedings 169–86.
56 Amalendu De, 'Raja Subodh Chandra Mallik and his Times', *Bengal Past and Present*, vol. 98, part 2, no. 187, p. 50.
57 Ibid., p. 51.
58 Ibid., pp. 52–5.
59 Circular no. 40–2, T. C., 8 November 1905, Home Dept. Public A, June 1906, Proceedings 169–86.
60 Ch. Sec. Govt. of E.B. & A, to Sec. to Govt. of India, 21 November 1905, Home Dept. Public A, June 1906, Proceedings 169–86.
61 Fuller, op.cit., pp. 131–2.
62 Ibid., p. 140.
63 Ibid., p. 128.

64 Mary to her mother, 10 November [1905], MCP.
65 Ibid.
66 King Edward VII to the Prince of Wales, as quoted by Kenneth Rose, *King George V* (London: 1969), p. 62.
67 Calcutta, 7 February 1900, *Indian Speeches*, 4 vols. (Calcutta: 1900–6), vol. 3, pp. 221–2.
68 See, *Pioneer*, 10–18 November 1905.
69 Lawrence to Curzon, 22 February 1906, CA 183.
70 Mary to her mother, 31 December [1905], MCP.
71 Sir H. Campbell Bannerman (1836–1908), Prime Minister of England 1905–8. Viscount John Morley (1838–1923), Secretary of State for India and with Viceroy Minto, author of the Morley–Minto Reforms of 1909.
72 Curzon to Elgin, 17 November 1898, CA 182.
73 Kenneth Rose, *Superior Person* (London: 1969), p. 367.
74 Mary to her mother, 25 October [1905], MCP.
75 The King-Emperor to Curzon, 15 September 1905, CP vol. 136.
76 Lawrence to Curzon, 30 September 1905, CA 183.
77 Curzon to Mrs Leiter, 24 October 1905, MCP.
78 Kitchener to Marker, 21 September 1905, KMP.
79 Godley to Lawrence, 8 October 1905, WLP vol. 42.
80 Curzon to Ampthill, 12 August 1905, AP vol. 19.
81 Mary to her parents, [August 1905], MCP.
82 Curzon 'Notes on Early Life', CA 20.
83 Brodrick to Curzon, 14 December 1898, CP vol. 10.
84 The Earl of Middleton, *Records and Reactions, 1856–1939* (London: 1939), p. 18.
85 Brodrick to Curzon, 26 September 1901, CP vol. 10B.
86 Curzon to Mrs Leiter, 24 October 1905, MCP.
87 St. John Brodrick, *Relations of Lord Curzon as Viceroy of India with the British Government, 1902–5*, (London: 1926).
88 Ibid.
89 Nigel Nicolson, *Mary Curzon* (London: 1977), p. 184.
90 Mary to her mother, 9 January 1906, MCP.
91 Curzon to Mrs Leiter, 19 July 1906, MCP.

Bibliography

This book is largely based on private papers and the relevant despatches and proceedings in the different departments of the Government of India.

ARCHIVAL SOURCES AND PRIVATE PAPERS

BRITAIN

India Office Library and Records, London
Curzon Papers (MSS Eur.F.111)
Curzon Additional Papers
Walter Lawrence Papers (MSS Eur.F.143)
Government of India Records in Curzon's private papers: Official papers on Office Routine 1898–9, vol.239; Official papers on the Partition of Bengal with notes by Curzon 1903–5, vol.247; Proceedings of the Educational Conference with notes by Curzon 1901, vol.248; Report of the Indian University Commission 1902, vols.662 and 663; Viceroy's note on the Military Department in the 9th Lancer's Case 1902, vol. 402; Private Correspondence relating to Military Administration in India 1902–5, vol.412.

British Museum, London
Kitchener–Marker Papers (Add. MSS 52276–8)

Eastbourne Central Library
Oscar Browning Papers

Courtesy Lady Alexandra Metcalfe
Mary Curzon Papers, edited for publication by Prof. John Bradley and in the ownership of Lady Alexandra Metcalfe who kindly granted me access.

INDIA

National Archives of India, New Delhi
Ampthill Papers (MSS Eur.E.233) in 2 microfilm reels
Curzon Papers (MSS Eur.F.111) in 14 microfilm reels
Gokhale Papers in 4 microfilm reels
Government of India Records
Native Newspaper Reports 1899–1905
The relevant Educational, Municipal and Public Series of the Home Department and Secret and Political Series in the Military Department in Curzon's viceroyalty have been consulted with the specific references given in the notes.

NEWSPAPERS

Nehru Memorial Museum Library, New Delhi
Amrita Bazaar Patrika
Bande Mataram
Englishman
The Pioneer

National Library of India, Calcutta
The Statesman

W. H. Targett Archives, New Delhi
Capital

WORKS BY GEORGE NATHANIEL CURZON

'The Conservatism of Young Oxford'. *The National Review*, vol.3, 16, 1884.
'Peers' Disabilities'. Privately printed, London, 1894.
Problems of the Far East. London, 1896.
Indian Speeches. 4 vols. Calcutta, 1900–6.
'The Place of India in the Empire'. Privately printed, London, 1909.
Kedleston Church. Privately printed, London, 1922.
Tales of Travel. London, 1923.
British Government in India. 2 vols. London, 1925.
Leaves from a Viceroy's Note-Book. London, 1926.
Walmer Castle and its Lords Warden. London, 1927.

PUBLISHED SOURCES

Abbot, Evelyn and Lewis Campbell. *The Life and Letters of Benjamin, Jowett, M. A.* London, 1897.

Anstruther, Ian. *Oscar Browning*. London, 1983.

Aronson, Theo. *Victoria and Disraeli*. London, 1977.

Asquith, Margot. *The Autobiography of Margot Asquith*. London, 1920.

Ballhatchet, Kenneth. *Race, Sex and Class Under the Raj*. London, 1980.

Balsan, Consuelo Vanderbilt. *The Glitter and the Gold*. London, 1953.

Bamford, T. W. *Rise of the Public Schools*. London, 1967.

Banerjea, Sir Surendranath. *A Nation in the Making*. Calcutta, 1963.

Barnett, Corelli. *The Collapse of British Power 1919–1945*. London, 1972.

Barton, William. *The Princes of India*. London, 1934.

Basu, Aparna. *The Growth of Education and Political Development in India, 1898–1920*. Delhi, 1974.

Beames, John. *Memoirs of a Bengal Civilian*. London, 1961.

Bence-Jones, M. *Viceroys of India*. London, 1982.

Benson, Arthur. *The Myrtle Bough*. Eton, 1903.

Benson, E. F. *As We Were*. London, 1938.

Besant, Annie. *How India Wrought For Freedom*. Madras, 1915.

Bishui, Kalpana. 'The Origins and Evolution of the Scheme for the first Partition of Bengal', *Quarterly Review of Historical Studies*, vol. 5, 2 (1965–6).

Blake, Robert. *The Unknown Prime Minister: the Life and Times of Andrew Bonar Law*. London, 1955.

————. *Disraeli*. London, 1956.

Bott, Alan. *Our Fathers*. London, 1931.

Bradley, John. *Lady Curzon's India*. London, 1985.

Briggs, Asa. *Victorian People*. London, 1955.

Brodrick, St. John. *Relations of Lord Curzon as Viceroy of India with the British Government, 1902–5*. Privately printed, London, June 1926.

Browning, Oscar. *Memories of Sixty Years*. London, 1923.

Buck, Edward J. *Simla Past and Present*. Bombay, 1925.

Buckland, C. E. *Bengal Under the Lieutenant-Governors*. 2 Vols. Calcutta, 1901.

Butler, Iris. *The Viceroy's Wife*. London, 1969.

Chakrabarti, Hiren. *Political Protest in Bengal: Boycott and Terrorism 1905–1918*. Calcutta, 1992.

Chirol, Valentine. *Indian Unrest*. London, 1910.

Churchill, Sir W. *Great Contemporaries*. London, 1937.

Coleman, James C. *Abnormal Psychology and Modern Life*. Bombay, 1981.

Cowles, Virginia. *Edward VII and his Circle*. London, 1956.

Crewe, Quentine. *The Frontiers of Privilege*. London, 1961.

Cronin, R. P. *British Policy and Administration in Bengal.* Calcutta, 1977.

Cumpston, Mary. 'Some Early Indian Nationalists and their Allies in the British Parliament, 1815–1906', *English Historical Review,* vol.26 (April 1901).

Curzon of Kedleston, Marchioness. *Reminiscences.* London, 1955.

D'Abernon, Viscount. *An Ambassador of Peace.* London, 1929.

Davis, H. W. C. *A History of Balliol College.* London, 1963.

De, Amalendu. 'Raja Subodh Chandra Mallik and His Times', *Bengal Past and Present,* vol.98, part 2, 187 (July-December 1979).

Dilks, David. *Curzon in India.* 2 vols. London, 1969.

Diver, Maud. *Royal India.* London, 1942.

Dufferin and Marchioness Ava. *Our Viceregal Life in India.* 2 vols. London, 1890.

Dugdale, B. E. C. *Arthur James Balfour.* London, 1936.

Dutt, Palme. *India Today.* Bombay, 1949.

Dutt, R. C. *Open Letters to Lord Curzon.* Calcutta, 1904.

Eden, Emily. *Up the Country.* London, 1930.

Edwards, Michael. *High Noon of Empire.* London, 1965.

Egremont, Max. *The Cousins.* London, 1977.

Ellis, Havelock. *Studies in the Psychology of Sex.* 2 vols. New York, 1942.

Ensor, R. C. K. *England, 1870–1914.* Oxford, 1936.

Erikson, Erik H. *Identity: Youth and Crisis.* New York, 1968.

———. *Gandhi's Truth: On the Origins of Militant Nonviolence.* New York, 1969.

———. *Childhood and Society.* London, 1974.

Faber, Geoffrey. *Jowett.* London, 1957.

Fitzroy, Yvonne. *Courts and Camps in India.* London, 1926.

Fraser, Sir Andrew H. L. *Among Indian Rajahs and Ryots.* London, 1911.

———. *Civil Servants Recollections and Impression of Thirty-seven Years Work in the Central Provinces and Bengal.* London, 1912.

Fraser, Lovat. *India Under Curzon and After.* London, 1911.

Fuller, Bampfylde. *Some Personal Experiences.* London, 1930.

———. *The Empire of India.* London, 1913.

Gallagher, John. *Decline, Revival and Fall of the British Empire.* Cambridge, 1982.

Gambier-Parry, Ernest. *Annals of an Eton House.* London, 1907.

Gathorne-Hardy, Jonathan. *The Rise and Fall of the British Nanny.* London, 1972.

Ghosh, Aurobindo. *Bankim Chandra.* Pondicherry, 1954.

———. *Sri Aurobindo on Himself and on the Mother.* Pondicherry, 1953.

Ghosh, P. C. *The Development of the Indian National Congress 1892–1960.* Calcutta, 1960.

Gilbert, Martin. *Laby Randolph Churchill.* London, 1969.

Gilbert, Martin. ed. *Servant of India, Sir James Dunlop Smith.* London, 1966.

Glyn, Sir Anthony. *Elinor Glyn*. London, 1968.

Gopal, S. *British Policy in India 1858–1905*. Cambridge, 1965.

Hardy, Peter. *Muslims in British India*. Cambridge, 1972.

Hibbert, Christopher. *The Illustrated London News*. London, 1975.

———. *The Great Mutiny*. London, 1978.

Hobson, J. A. *Imperialism, A Study*. New York, 1902.

Horner, Frances. *Time Remembered*. London, 1933.

Hunter, W. W. *Life of the Earl of Mayo*. London, 1875.

Hutchings, Frances G. *The Illusion of Permanence*. Princeton, 1967.

Hyam, Ronald. *Britain's Imperial Century, 1815–1914*. London, 1976.

Joshi, V. G. ed. *Lala Lajpat Rai, Autobiographical Writings*. Delhi, 1965.

Karve, D. G. and D. V. Ambedkar. eds. *Speeches and Writings of Gopal Krishna*. 3 vols. Poona, 1966.

Kerr, James Campbell. *Political Troubles in India, 1907–17*. Calcutta, 1973.

Kiernan, V. G. *The Lords of Human Kind*. London, 1969.

King, Peter. ed. *A Viceroy's India: Leaves from Lord Curzon's Note-book*. London, 1984.

———. *The Viceroy's Fall*. London, 1986.

Kinkaid, Denis. *British Social Life in India 1608–1937*. London, 1973.

Lajpat, Rai. *The Arya Samaj*. Lahore, 1932.

Lawrence, Sir Walter. *The India We Served*. London, 1928.

Leslie, Sir Shane. *Studies in Sublime Failure*. London, 1932.

Lipsett, H. C. *Lord Curzon in India 1898–1903*. London, 1903.

Long, James. 'Echoes from Old Calcutta: Its Localities', *Calcutta Review*, vol.36 (1852).

Lord, John. *The Maharajas*. Delhi, 1971.

Low, D. A. *Lion Rampant: Essays in the Study of British Imperialism*. London, 1973.

Macmillan, Sir Harold. *Winds of Change*. London, 1966.

Maconochie, Sir Evan. *Life in the Indian Civil Service*. London, 1926.

Magnus, Sir Philip. *Kitchener, Portrait of an Imperialist*. London, 1958.

Majumdar, A. C. *Indian National Evolution*. Madras, 1917.

Majumdar, R. C. *History of the Freedom Movement in India*. 3 vols. Calcutta, 1971.

Malhotra, P. L. *The Administration of Lord Elgin 1894–99*. New Delhi, 1979.

Martineau, Harriet. *The British Rule in India*. London, 1857.

Massey, Montague. *Recollections of Calcutta for Over Half a Century*. Calcutta, 1918.

Mclane, John R. 'The Decision to Partition Bengal in 1905', *The Indian Economic and Social History Review*, vol. 2, 3 (July 1965).

———. *Indian Nationalism and the Early Congress*. Princeton, 1977.

Mehra, P. *The Younghusband Expedition*. Bombay, 1968.

Menpes, Mortimer. *The Durbar*. London, 1903.

Metcalfe, Thomas R. *The Aftermath of Revolt, India 1857–1870*. Princeton, 1964.

Middlemas, K. and J. Barnes. *Baldwin*. London, 1969.

Middleton, Earl of. *Records and Reactions 1856–1939*. London, 1939.

Mosley, Leonard. *Curzon, the End of an Epoch*. London, 1960.

Mosley, Oswald. *My Life*. London, 1968.

Mukherjee, Haridas and Uma Mukherjee. *India's Fight for Freedom or the Swadeshi Movement 1905–6*. Calcutta, 1958.

———. *Sri Aurobindo and the New Thought in Indian Politics*. Calcutta, 1964.

Neville, Ralph. *Leaves from the Notebooks of Lady Dorothy Neville*. London, 1907.

Nevinson, H. *The New Spirit in India*. London, 1908.

Nicolson, Sir Harold. *Curzon, the Last Phase, 1919–1925*. London, 1934.

———. *Some People*. London, 1947.

Nicolson, Nigel. *Mary Curzon*. London, 1977.

O'Donnell, Charles. *The Cause of Present Discontent in India*. London, 1908.

Pal, Bipin Chandra. *Memories of My Life and Times*. 2 vols. Calcutta, 1932, 1951.

———. *Character Sketches*. Calcutta, 1957.

Pike, E. Royston. ed. *Britain's Prime Ministers*. Middlesex, 1968.

Pullar, Philippa. *Gilded Butterflies*. London, 1978.

Ranade, M. G. *The Miscellaneous Writings*. Bombay, 1915.

Ravensdale, Baroness. *Little Innocents: Childhood Reminiscences*. London, 1932.

Ray, Prithwischandra. *The Case Against the Break-up of Bengal*. Calcutta, 1905.

Ray, Rajat Kanta. *Social Conflict and Political Unrest in Bengal, 1875–1927*. Delhi, 1984.

Ribbelsdale, T. L. *Impressions and Memories*. London, 1927.

Riddell, Lord. *Lord Riddell's Intimate Diary of the Peace Conference and After*. London, 1933.

———. *More Pages from My Diary, 1908–1914*. London, 1934.

Rodd, Rennell. *Diplomacy*. 3 vols. London, 1929.

Ronaldshay, Earl of. *The Life of Lord Curzon*. 3 vols. London, 1928.

Rose, Kenneth. *Superior Person*. London, 1969.

——— *King George V*. London, 1979.

Rumbold, Algernon. *Watershed in India: 1914–1922*. London, 1979.

Sarkar, Sumit. *The Swadeshi Movement in Bengal 1903–8*. New Delhi, 1977.

Sarkar, Susobhan. *Bengal Renaissance*. New Delhi, 1970.

Seal, Anil. *The Emergence of Indian Nationalism*. Cambridge, 1968.

Searle, G. R. *The Quest for National Efficiency*. Oxford, 1971.

Seeley, J. R. *The Expansion of England*. London, 1899.

Sinha, Pradip. *Calcutta in Urban Hisotry*. Calcutta, 1978.

Stephen, Sir James Fitzjames. *Liberty, Equality, Fraternity*. London, 1872.

———. 'Foundations of the Government of India', *Nineteenth Century*, vol.80 (October 1883).

Stephen, Leslie. *The Life of Sir James Fitzjames Stephen*. New York, 1895.

Stokes, Eric. *The English Utilitarians and India*. Oxford, 1959.

Stone, Lawrence. *The Family, Sex and Marriage in England, 1500–1800*. London, 1979.

Strachey, Sir John. *India, Its Administration and Progress*. London, 1888.

Tilak, B. G. *Bal Gangadhar Tilak: His Writings and Speeches*. Madras, 1918.

Trevelyan, G. O. *The Life and Letters of Lord Macaulay*. 2 vols. New York, 1876.

Tripathi, Amales. *The Extremist Challenge*, Calcutta, 1962.

Vansittart, Lord. *The Mist Procession*. London, 1958.

Waterhouse, Nourah. *Private and Official*. London, 1942.

Wedderburne, W. *Alan Octavius Hume*. London, 1913.

Wilson, Robert. *The Life and Times of Queen Victoria*. London, 1887.

Wolman, Benjamin B. ed. *The Psychoanalytic Interpretation of History*. New York, 1971.

Wolpert, Stanley A. *Tilak and Gokhale*. Los Angeles, 1962.

Woodruff, Philip. *The Men who Ruled India*. 2 vols. New York, 1934.

Wortham, H. E. *Victorian England and Cambridge*. London, 1956.

Young, G. M. ed. *Victorian England*. London, 1953.

Young, Wayland. *Eros Denied*. London, 1964.

Index